Synthesis Lectures on Green Energy and Technology

This series publishes short books in environmental-friendly and sustainable technologies, in particular state-of-the-art research and development in green energy and pollution control. The books in this series are written by top researchers and professionals. The series addresses researchers, advanced students, technical consultants as well as decision makers in industries and politics. Hence, the level of presentation spans from instructional to highly technical.

Lei Wang · Zhongchao Tan · Yimin Wu

Modulation Strategies of Cu-based Electrocatalysts for Enhancing Electrocatalytic CO_2 Conversion

Springer

Lei Wang
Department of Mechanical and Mechatronics Engineering
University of Waterloo
Waterloo, ON, Canada

Yimin Wu
Department of Mechanical and Mechatronics Engineering
University of Waterloo
Waterloo, ON, Canada

Zhongchao Tan
Vice Provost and Dean of Academic Affairs
Eastern Institute of Technology
Ningbo, Zhejiang, China

ISSN 2948-2720 ISSN 2948-2739 (electronic)
Synthesis Lectures on Green Energy and Technology
ISBN 978-3-031-98055-8 ISBN 978-3-031-98056-5 (eBook)
https://doi.org/10.1007/978-3-031-98056-5

This Springer imprint is published by the registered company Springer Nature Switzerland AG
The registered company address is: Gewerbestrasse 11, 6330 Cham, Switzerland

Acknowledgements The inspiration for this book stems from my doctoral research. As I delved deeper into my field, constantly exploring possible solutions, I was also enriching myself and expanding my knowledge. The original intention of writing this book was to share my research perspectives and the progress in this field with a broader audience.

Most importantly, I would like to express my sincere gratitude to my supervisors—Prof. Yimin Wu and Zhongchao Tan from the University of Waterloo. Their patient guidance and rigorous academic attitude have greatly benefited me. The research accomplishments presented in this book were made possible thanks to their invaluable support and mentorship, which enabled me to tackle every research challenge to the best of my ability.

Throughout the writing process, I encountered various difficulties and challenges. However, the companionship and support of my family gave me immense encouragement, allowing me to persevere and complete this book. Because of their unconditional love, I have never felt alone.

Ultimately, completing this book is not an end, but a new beginning. Along the way, I have learned many valuable lessons—about research, perseverance, and the meaning of the future. I hope to share these reflections with readers as well.

Competing Interests The authors have no competing interests to declare that are relevant to the content of this manuscript.

Contents

Abbreviations

ACB	Activated carbon black
AC-HAADF-STEM	Aberration-corrected high-angle annular dark-field scanning transmission electron microscope
AEM	Anion exchange membrane
ATR FT-IR	Attenuated total reflection Fourier transform infrared spectroscopy
CV	Cyclic voltammetry
DFT	Density functional theory
DMPO	5,5-dimethyl-1-pyrroline N-oxide
ECR	Electrochemical CO_2 reduction
ECSA	Electrochemical surface area
EDS	Energy dispersive X-ray spectroscopy
EELS	Electron energy loss spectroscopy
EIS	Electrochemical impedance spectroscopy
EPR	Electron paramagnetic resonance
EXAFS	Extended X-ray absorption fine structure
FE	Faradaic efficiency
GBO	Grain boundary oxidation
GBs	Grain boundaries
GDEs	Gas diffusion electrode
GDL	Gas diffusion layer
GIXRD	Grazing incidence X-ray diffraction
HER	Hydrogen evolution reaction
HRTEM	High-resolution transmission electron microscopy
MEA	Membrane electrode assembly
OCP	Open-circuit potential
OD-Cu	Oxide-derived copper
OER	Oxygen evolution reaction
PCET	Proton-coupled electron transfer

PDOS	The projected density of states
PTFE	Polytetrafluoroethylene
RDS	Rate-determining step
RHE	Reversible hydrogen electrode
SEM	Scanning electron microscope
XANES	X-ray absorption near-edge structure
XAS	X-ray absorption spectroscopy
XPS	X-ray photoelectron spectroscopy

List of Figures

List of Tables

Introduction

1

1.1 Background

Since the Industrial Revolution, CO_2 emissions have surged dramatically and continue to rise [1]. CO_2 is not only produced by fossil fuels combustion but is also a natural byproduct of respiration. As main greenhouse gas, CO_2 has contributed significantly to global warming, climate change, and ocean acidification. In areas near factories, elevated CO_2 concentrations can lead to severe health problems [2–4]. As a signatory of the Paris Climate Agreement, Canada has committed to reducing its greenhouse gas emissions by 30% below 2005 levels by 2030. Although Canada accounts for 1.6% of global emissions, the country aim to achieve net-zero carbon emissions by 2050 [5, 6]. To reach carbon neutrality, there is an urgent need to develop novel and efficient CO_2 conversion technologies (Fig. 1.1).

To close the carbon cycle, one potential approach is to transition the energy and chemical industries' reliance from fossil fuels to renewable sources such as wind and solar power [8, 9]. However, renewable energy currently accounts for only 30% of the total energy supply, which is insufficient to meet industrial demand without the development of large-scale energy storage technologies [10]. An alternative approach is to capture CO_2 emissions from factories and convert them into new chemicals and fuels, which may be more practical [11]. Electrocatalytic CO_2 reduction (ECR) to high-value chemicals and fuels using renewable energy offers a promising method to mitigate CO_2 emissions and achieve carbon neutrality. As illustrated in Fig. 1.2, renewable energy powers the CO_2 electrolyzer, which converts CO_2 and water into chemicals and fuels. These fuels can be stored a long time and transported to industries in need. The emitted CO_2 can be recaptured and reused as a reactant, thus closing the carbon cycle. Additionally, C_{2+} products like ethylene can serve as raw materials for more complex chemical syntheses [12, 13].

L. Wang et al., *Modulation Strategies of Cu-based Electrocatalysts for Enhancing Electrocatalytic CO_2 Conversion*, Synthesis Lectures on Green Energy and Technology,
https://doi.org/10.1007/978-3-031-98056-5_1

Fig. 1.1 A diagram for net-zero carbon emissions [7]. Reprinted with permission. Copyright 2023 Springer Nature

However, CO_2 is a stable linear molecule with strong C=O bonds, making its electrochemical reduction challenging [14]. Overcoming the reaction energy barrier of ECR requires a significant overpotential [15]. Moreover, ECR involves multiple proton-coupled electron transfer (PCET) processes, which complicates the reaction pathway [16, 17]. Considering the diversity of ECR products, such as CO, formic acid, methane, ethylene, ethanol, and acetic acid, achieving highly efficient and selective product generation is a significant challenge. Therefore, rational catalyst design strategies and a deep understanding of the catalytic mechanisms are crucial for directing ECR towards specific products.

1.2 Objectives

This book aims to provide some guidelines for highly efficient electrocatalysts for enhancing CO_2 conversion. The initial focus of the experimental research is to explore catalytic mechanisms in lab-scale H-cells using a neutral electrolyte. In situ multimodal characterizations reveal that oxidized $Cu^{\delta+}$ species promote the generation of C_{2+} products while nanoalloy effects favor methane formation in complex grain boundary assemblies. Leveraging the understanding of the relationship between active sites and ECR selectivity, activated carbon black was introduced in industrial-scale membrane electrode assembly (MEA) reactors to modulate the local microenvironment of the catalyst surface, stabilizing $Cu^{\delta+}$ species and promoting selective ethylene production at high current densities.

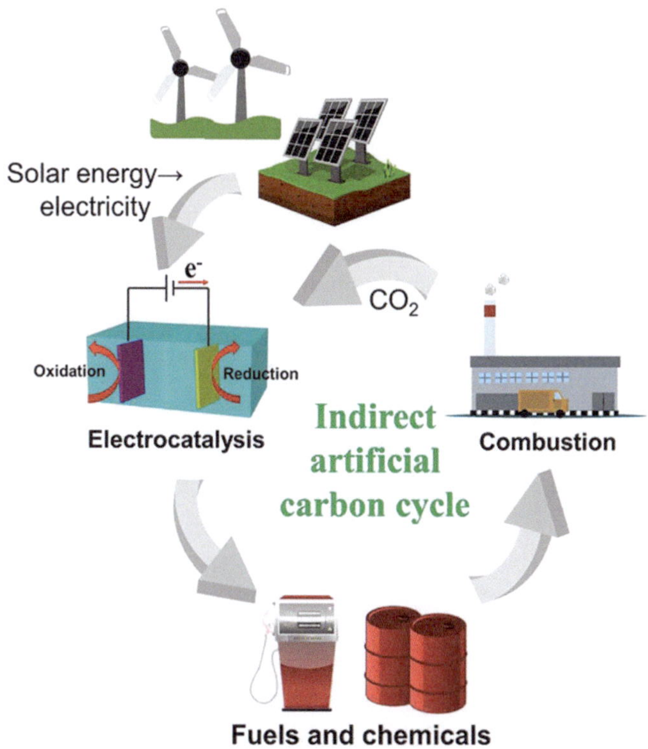

Fig. 1.2 Schematic illustrations of indirect artificial carbon cycles based on CO_2 conversions [13]. Reprinted with permission. Copyright 2022 American Chemical Society

This advancement accelerates the commercialization of ECR technology. The research program consists of the following contents:

1. We demonstrated an electro-shock reduction strategy to break through the immiscible limitation. The controllable grain boundaries (GBs) for the oxidization of $Cu^{\delta+}$ species by modulating Ag loading were developed. The Ag enrichment at the GBs facilitates the bonding of oxygen with the uncoordinated Cu atoms, resulting in grain boundaries oxidation (GBO) effect. The obtained CH_4 selectivity is twice that of GBs effect under neutral electrolyte in CuAg system. The enhanced performance is attributed to the stable $Cu^{\delta+}$ species, unique electron transfer mechanism, and strong tensile Cu–O bond strain from Cu–O–Ag structure.
2. We engineered four different GB structures to elucidate the real active sites and their dynamic evolution processes within complex GBs assembles. Multiple in situ characterizations revealed that GBs and GBO strengthen resistance of the oxidative $Cu^{\delta+}$ species to the electrochemical reduction. The in situ generated strongly oxidative

hydroxyl radicals altered local reaction environment on the catalyst surface, inducing and stabilizing oxidative $Cu^{\delta+}$ species. The oxidation state of copper plays a crucial role in the C_{2+} product pathway, while the nanoalloy effect tends to the formation of CH_4 product.

3. The functionalized activated carbon black (ACB) was introduced to modulate the local microenvironment of Cu NPs in the pH-neutral electrolyte. The enhanced performance of ECR to C_2H_4 (55.6% ± 2.8 FE at 316 mA cm^{-2}) in an MEA reactor was attributed to the locally oxidative microenvironment on the catalyst surface, regulated by hydroxyl species generated in situ from water dissociation. The well-stabilized $Cu^{\delta+}$-OH species served as active sites during MEA testing, leading to an irreversible and asynchronous change in morphology and valence, yielding high-curvature nanowhiskers.

1.3 Overview of the Book

This book focuses on establishing the relationship between the real active sites of catalysts and ECR selectivity under neutral conditions, revealing the dynamic evolution of active sites under ECR reaction conditions. This will help guide the rational design of catalysts and advance the industrial application of ECR technology.

This Chapter is an introduction of this research program, covering research motivations, objects and overview of the book.

Chapter 2 covers the literature review of modulation strategies of Cu-based electrocatalysts, CO_2 electrolyzer and gas diffusion electrodes for ECR.

Chapter 3 developed a novel electro-shock method for precise control of catalysts with GBs and GBO structures. 1% Ag loading in Cu matrix led to the Ag enrichment on the GBs, triggering the oxidation of Cu sites (i.e., GBO effect). In situ XAS analysis reveals that, unlike GBs effect, the performance improvement originates from the steady-state $Cu^{\delta+}$ species, a unique electron transfer mechanism, and a strong tensile strain induced by the GBO effect in CuAg system. Overall, this work not only establishes the relationship between GBO effect and CO_2-to-CH_4 conversion in bimetallic CuAg systems but also provides precise elucidation of active sites derived from ternary Cu–O–Ag system.

Chapter 4 focused on commonly used grain boundaries (GBs), oxidation states, and alloying strategies, constructing four different types of catalysts structures: original Cu GBs, oxygen-enriched grain boundary oxidation (GBO), Ag-enriched GBO, and Cu/Ag GBs. Catalytic activity comparisons indicate that the oxidation state of Cu plays a decisive role in the CO_2-to-C_{2+} conversion, and nanoalloy effect tends to favor the CH_4 production in intricate GBs assemblies. This study clarifies the impact of each performance modulation strategy on catalytic activity, aiding in the avoidance of blindly combining multiple design strategies.

Chapter 5 presented a strategy that manipulates the interfacial microenvironment of Cu nanoparticles in an MEA to suppress hydrogen evolution reaction and enhance C_2H_4 conversion. In situ multimodal characterizations consistently reveal well-stabilized $Cu^{\delta+}$-OH species as active sites during MEA testing. The OH radicals generated in situ from water create a locally oxidative microenvironment on the copper surface, stabilizing the $Cu^{\delta+}$ species and leading to an irreversible and asynchronous change in morphology and valence, yielding high-curvature nanowhiskers. Understanding the reaction mechanism in a MEA electrolyzer would better guide the redesign and assembly of catalysts on the GDEs within MEA, accelerating the commercialization of ECR.

Chapter 6 gives the summary and discussion of the completed research and future works based on current progresses.

Competing Interests The authors have no conflicts of interests to declare that are relevant to the content of this chapter.

References

1. Friedrich, J., Damassa, T.: The history of carbon dioxide emissions (2014)
2. Solomon, S., et al.: Irreversible climate change due to carbon dioxide emissions. Proc. Natl. Acad. Sci. U.S.A. **106**(6), 1704–1709 (2009)
3. Gattuso, J.-P., et al.: Contrasting futures for ocean and society from different anthropogenic CO_2 emissions scenarios. Science **349**(6243), aac4722 (2015)
4. Yoro, K.O., Daramola, M.O.: CO_2 emission sources, greenhouse gases, and the global warming effect. In: Advances in Carbon Capture, pp 3–28. Elsevier (2020)
5. Environment; Canada, C. C.: National inventory report 1990–2018: greenhouse gas sources and sinks in Canada. Environment Canada Ottawa (2018)
6. Donner, S., Zickfield, K.: Canada's contribution to meeting the temperature limits in the Paris climate agreement (2016)
7. Eldesouki, M.H., et al.: A comprehensive overview of carbon dioxide, including emission sources, capture technologies, and the conversion into value-added products. Clean Technol. Environ. Policy **25**(10), 3131–3148 (2023)
8. Armstrong, R., et al.: The frontiers of energy. Nat. Energy **1**(15020), 2015 (2016)
9. Obama, B.: The irreversible momentum of clean energy. Science **355**(6321), 126–129 (2017)
10. Ma, S., Kenis, P.J.: Electrochemical conversion of CO_2 to useful chemicals: current status, remaining challenges, and future opportunities. Curr. Opin. Chem. Eng. **2**(2), 191–199 (2013)
11. De Luna, P., et al.: What would it take for renewably powered electrosynthesis to displace petrochemical processes? Science **364**(6438), eaav3506 (2019)
12. Yang, H., et al.: A review of the catalytic hydrogenation of carbon dioxide into value-added hydrocarbons. Catal. Sci. Technol. **7**(20), 4580–4598 (2017)
13. Low, J., et al.: Identification and design of active sites on photocatalysts for the direct artificial carbon cycle. Acc. Mater. Res. **3**(3), 331–342 (2021)
14. Li, K., et al.: Recent advances in heterogeneous photocatalytic CO_2 conversion to solar fuels. ACS Catal. **6**(11), 7485–7527 (2016)

15. Verma, S., et al.: A gross-margin model for defining technoeconomic benchmarks in the electroreduction of CO_2. Chemsuschem **9**(15), 1972–1979 (2016)
16. Kortlever, R., et al.: Catalysts and reaction pathways for the electrochemical reduction of carbon dioxide. J. Phys. Chem. Lett. **6**(20), 4073–4082 (2015)
17. Gattrell, M., et al.: A review of the aqueous electrochemical reduction of CO_2 to hydrocarbons at copper. J. Electroanal. Chem. **594**(1), 1–19 (2006)

Literature Review 2

2.1 Background

Over the past few decades, rising CO_2 emissions from fossil fuel combustion have posed potential threats to the global climate, such as ocean acidification, extreme weather events, and droughts. High concentrations of CO_2 can also lead to health risks and water security challenges [1–3]. The rapid increase in CO_2 emissions has prompted scientists and governments to collaborate on achieving net-zero emissions and developing efficient CO_2 conversion technologies [3, 4]. Electrocatalytic CO_2 reduction (ECR) into highly value-added chemicals and fuels using renewable energy is a promising method to reduce carbon emissions and support the development of a sustainable society [5].

A typical CO_2 electrolyzer consists of a cathode chamber and an anode chamber separated by an ion exchange membrane. The ECR reaction occurs at the cathode, while the oxygen evolution reaction (OER) usually occurs at the anode (Fig. 2.1) [6]. CO_2, being a fully oxidized and thermodynamically stable molecule, poses several significant challenges for electrochemical conversion. Firstly, ECR is a thermodynamically uphill reaction with a high activation barrier, necessitating a sufficient overpotential to drive the reaction and make CO_2 conversion feasible [7]. Over 16 products have been reported from ECR. Table 2.1 provides the equilibrium potentials required for commonly reported electrochemical reduction products [8]. Secondly, ECR involves multiple electron/proton transfer steps, with electrons and protons supplied by water [9]. This proton-assisted process involves various intermediates and products, complicating the control of product selectivity [10]. In aqueous electrolytes, the competing hydrogen evolution reaction (HER) can inhibit the formation of reduced products, further complicating ECR [11]. Therefore, achieving effective ECR relies on the rational design of catalysts to minimize side reactions and enhance selective production of reduced products.

L. Wang et al., *Modulation Strategies of Cu-based Electrocatalysts for Enhancing Electrocatalytic CO_2 Conversion*, Synthesis Lectures on Green Energy and Technology,
https://doi.org/10.1007/978-3-031-98056-5_2

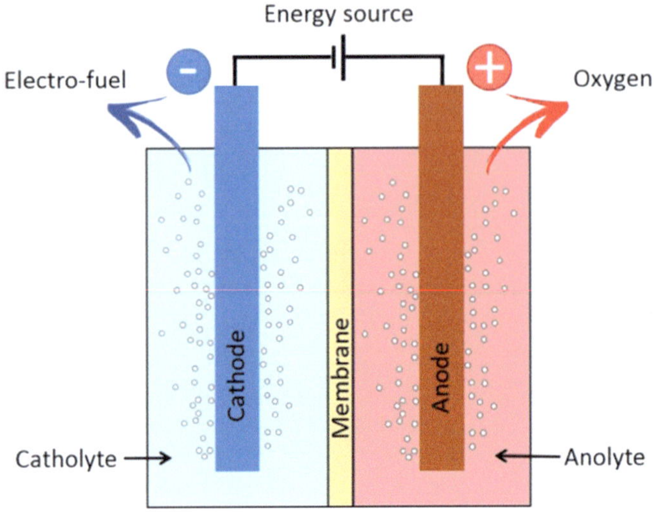

Fig. 2.1 A typical schematic of an ion-conducting membrane based CO_2 electrolyzer [6]. Reprinted with permission. Copyright 2016 Elsevier

Table 2.1 ECR reaction and thermodynamic potentials [8]

Product	ECR reaction	E^0 (V vs. RHE)
Carbon monoxide	$CO_2(g) + H_2O(l) + 2e^- \rightarrow CO(g) + 2OH^-(aq)$	– 0.10
Formate	$CO_2(g) + H_2O(l) + 2e^- \rightarrow HCOO^-(aq) + OH^-(aq)$	– 0.20 (for pH < 4); – 0.20 + 0.059[pH-4] (for pH > 4)
Methanol	$CO_2(g) + 5H_2O(l) + 6e^- \rightarrow CH_3OH(aq) + 6OH^-(aq)$	0.03
Methane	$CO_2(g) + 6H_2O(l) + 8e^- \rightarrow CH_4(g) + 8OH^-(aq)$	0.17
Acetate	$2CO_2(g) + 5H_2O(l) + 8e^- \rightarrow CH_3COO\text{-}(aq) + 7OH^-(aq)$	0.11
Ethylene	$2CO_2(g) + 8H_2O(l) + 12e^- \rightarrow C_2H_4(g) + 12OH^-(aq)$	0.08
Ethanol	$2CO_2(g) + 9H_2O(l) + 12e^- \rightarrow C_2H_5OH(aq) + 12OH^-(aq)$	0.09
n-propanol	$3CO_2(g) + 13H_2O(l) + 18e^- \rightarrow C_3H_7OH(aq) + 18OH^-(aq)$	0.10

Significant progress has been made in ECR research to date. Early ECR studies began in the 1950s [12], and in 1985, Yoshio Hori and co-workers quantified both gaseous and liquid ECR products using constant current tests on various polycrystalline metal electrodes in CO_2-saturated 0.5 M $KHCO_3$ electrolyte [13]. This was followed by systematic studies of ECR electrocatalysts. Research classified electrocatalysts into three categories based on their type: metals, non-metals, and molecular catalysts [14, 15]. Additionally, they can be categorized based on their binding energy to key CO_2 intermediates and HER intermediates, such as *H, *OCHO (bound to the surface through O), *COOH (bound to surface through C) and *CO (Fig. 2.2) [16]. For example, Pb, Hg, Tl, In, Sn, Cd and Bi primarily produce formic acid; Au, Ag, Zn, Pd and Ga mainly produce CO; and other metals such as Co, Ni, Fe, Ir, and Pt tend to facilitate HER. Among single-metal catalysts, only Cu efficiently reduce CO_2 to hydrocarbons and alcohols, such as methane, ethylene, ethanol, and acetate. The unique ECR performance of Cu originates from its negative adsorption energy for *CO and positive adsorption energy for *H [17, 18]. Therefore, to enhance the ECR conversion to multi-carbon products, more attention should be placed on modulation strategies for copper-based catalysts and the exploration of catalytic reaction mechanisms.

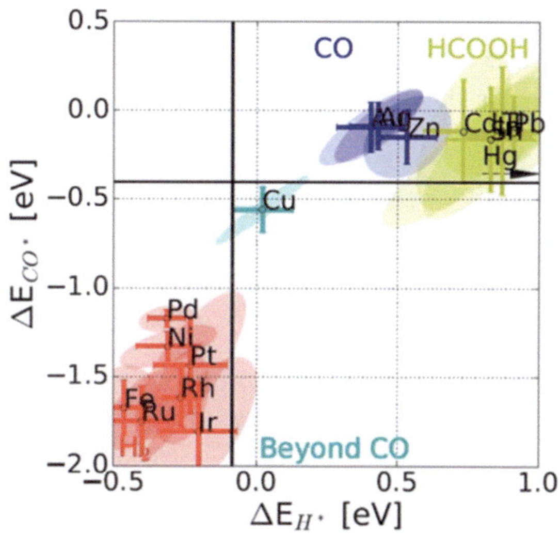

Fig. 2.2 CO_2 reduction metal classification based on the binding energy to key CO_2 reduction and HER intermediates [16]. Reprinted with permission. Copyright 2017 WILEY-VCH Verlag GmbH & Co. KGaA, Weinheim

2.2 Cu-Based Electrocatalysts

Among the state-of-the-art ECR catalysts, copper is the only metal capable of reducing CO_2 into hydrocarbons and alcohols [17]. Therefore, numerous approaches, including alloying [19, 20], oxide-derived copper (OD-Cu) [21, 22], grain boundaries (GBs) [23, 24], and morphology modification [25, 26], have been developed to increase the capability of the Cu-based catalysts. The main technique of these approaches is to create specific ensembles with controllable selectivity towards the targeted products (e.g. ethylene and methane). The revealed reaction mechanism and catalytic behaviors may further shine a light for development of Cu-based electrocatalysts.

2.2.1 Alloy Effect

Single-metal catalysts often produce only C_1 (e.g., CO, HCOOH) products (e.g., Au, Sn and Bi) and suffer from undesirable HER (e.g., Pt and Pd), or exhibit low product selectivity due to the diverse reaction pathways of Cu [27–29]. Consequently, many researchers have focused on alloying other metals with Cu. This approach aims to optimize the electronic structure and local coordination environment of Cu atoms, alter the adsorption strength of key intermediates, and expose new active sites, thereby enhancing the formation of multi-carbon products [30, 31]. Additionally, bimetallic catalysts can enhance CO coverage on the catalyst surface, promoting C–C coupling towards C_{2+} products (e.g., C_2H_4, and C_2H_5OH) [29]. Therefore, alloying strategies play a crucial role in modulating the reaction pathways of ECR [32].

The alloying effect on the formation mechanisms of C_1 and C_{2+} products is complex. Under severe cathodic conditions (i.e., cathodic potentials > − 1.1 V vs. reversible hydrogen electrode, RHE), CH_4 production tend to outcompete C_2H_4. This is because CH_4 formation requires a higher overpotential and faster kinetics [33]. In the CuAg alloy system, many studies reported CH_4 as the main reduced product. For instance, studies have shown that Ag–Cu surfaces, due to low-coordination sites, strongly bind with *CO, inhibiting CO–CO coupling and resulting in lower selectivity towards C_2H_4 (Fig. 2.3a, b) [34]. Ag-modified Cu nanowires ($Cu_{68}Ag_{32}$) exhibit more than three times the CH_4 activity and selectivity compared to pristine Cu nanowires. In situ characterizations reveal that the dynamic reoxidation/reduction-driven atomic interdiffusion of Cu sites facilitates the CuAg alloying (Fig. 2.3c). The induced tensile strain strengthens the binding with *CHO intermediates, promoting further reduction and suppressing undesired HER [35]. Furthermore, as both Cu and Ag have valence d-states, the catalytic performance of CuAg catalysts is in principle determined by their electronic structure. The interface interaction between Cu and Ag alters the surface electronic structure and tailors the adsorption strength of *CO on the catalytic surface (Fig. 2.3d). Consequently, *CO tends to convert into CH_4 before desorbing from the Cu surface [36]. Conversely, for the CuAg alloy system that produces C_{2+} products, the CO_2 reduction pathway follows a tandem catalysis mechanism. Ag sites are more favorable for CO production, which enriches CO at neighboring Cu sites and promotes C–C coupling towards C_{2+} products such as ethylene and ethanol [37]. For example, Gewirth and co-workers reported that a CuAg alloy prepared by additive-controlled electrodeposition achieves 60% ethylene selectivity and 25% ethanol selectivity (Fig. 2.3e). The high selectivity for C_{2+} products results from the stabilization of the Cu_2O overlayer and optimization of CO intermediates availability at Cu sites through Ag doping in the alloy system [19]. Zheng and co-workers reported that the introducing Ag facilitates electron transfer from Cu to Ag, creating electron-deficient Cu sites. This enhances the adsorption of key intermediates in the ethanol reaction pathway, such as $^*CH_3CHO$ and $^*CH_3CH_2O$, resulting in excellent ethanol selectivity (Fig. 2.3f) [38].

Additionally, the mixing modes of bimetallic catalysts influence the product distribution in ECR. Kenis and co-workers demonstrated that phase-separated PdCu catalysts achieve the highest ethylene selectivity (~50%) compared to both ordered and disordered counterparts [39]. The lower d-band center of these phase-separated catalysts, compared to Cu NPs, indicates the weaker CO intermediate binding, This suggests that geometric and structural effects play a more significant role than electronic effects in enhancing product distribution [39]. Carlos and colleagues reported that gold nanoparticles on polycrystalline copper foil follow a tandem catalysis mechanism, showing superior synergistic catalytic activity and selectivity compared to gold, copper, and CuAg alloys [40]. In summary, alloying strategies open new avenues for enhancing the conversion of ECR to target products, but different structures and interfaces will give different electronic states for different products selectivity.

2.2.2 Oxidation States

Beyond alloying strategies, modulation of the oxidation state also affects the activity and selectivity of ECR. For oxide-derived Cu (OD-Cu), the oxidation state of the metal centers is effectively maintained by residual oxygen species on the surface or subsurface of the catalyst after in situ reduction. These oxidized Cu species serve as active sites for ECR, facilitating C–C coupling to form C_{2+} products. For example, plasma-oxidized Cu can achieve ethylene selectivity exceeding 60% [41]. Comparative experiments with various plasma treatments confirm the crucial role of Cu^+ in the conversion of ECR to C_2H_4 (Fig. 2.4a) [41]. Edward and colleagues used in situ soft X-ray absorption spectroscopy to monitor the oxidation state of copper over time under ECR conditions. The sol–gel materials effectively slow the electrochemical reduction of oxidized copper species, stabilizing Cu^+ species under cathodic potentials (Fig. 2.4b) [42]. The catalyst achieves an ethylene partial current density of 160 mA cm^{-2} and an ethylene-to-methane ratio of 200 (Fig. 2.4c). However, after one hour of continuous operation, only 23% of Cu^+ species

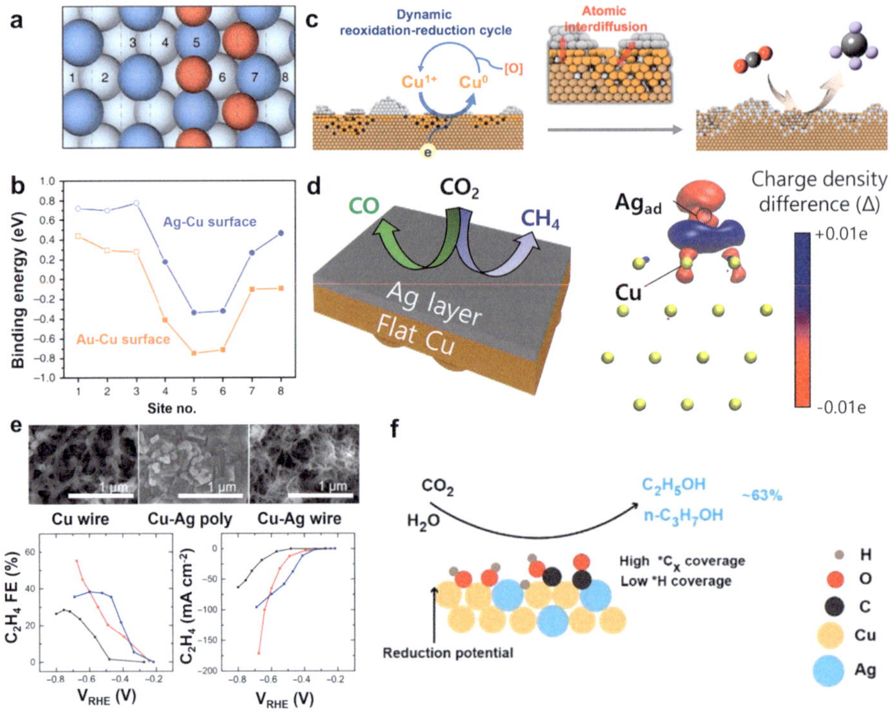

Fig. 2.3 Alloy effect of electrochemical CO_2 reduction. **a** Top view of the unit cell used for computational investigations (blue: top layer Ag or Au atoms; light blue: bottom layer Ag or Au atoms; orange: Cu atoms). The site numbers show the position and chemical environment of the binding sites [34]. Reprinted with permission. Copyright 2019 Springer Nature. **b** Energy diagram for CO adsorption on the Ag–Cu surface (blue line) and Au–Cu surface (orange line) [34]. Reprinted with permission. Copyright 2019 Springer Nature. **c** The dynamic reoxidation/reduction-driven atomic interdiffusion processes of Cu sites [35]. Reprinted with permission. Copyright 2020 American Chemical Society. **d** The optimized Cu/Ag layered catalyst for ECR-to-CH_4 conversion [36]. Reprinted with permission. Copyright 2020 Elsevier. **e** The scanning electron microscopy (SEM) images and corresponding C_2H_4 FE of CuAg bimetallic electrocatalysts from additive-controlled electrodeposition [19]. Reprinted with permission. Copyright 2018 American Chemical Society. **f** Illustration of the ECR process on the Cu_3Ag_1 catalyst [38]. Reprinted with permission. Copyright 2020 WILEY–VCH Verlag GmbH & Co. KGaA, Weinheim

can be effectively preserved (Fig. 2.4b). Theoretical analyses have further explained why OD-Cu leads to the production of ethylene and ethanol. Research indicates that 4-(6-) coordinated Cu adatoms and $Cu_3^{\delta+}O_3$ leads to tethering CO_2, while metastable near-surface oxygen atoms in fcc-(111) or (100)-like Cu matrix promote C–C coupling, thereby enhancing the formation of C_{2+} products [43]. Nevertheless, the poor stability of oxidized $Cu^{\delta+}$ species under high current densities poses significant challenges for the widespread application of OD-Cu. To enhance the stability of oxidized $Cu^{\delta+}$ species, Fang and colleagues developed a one-step surface coordination method to synthesize ultrastable and hydrophobic $Cu^{\delta+}$, achieving a C_2 Faradaic efficiency (FE) of 90.6% at a partial current density of 453.3 mA cm^{-2} (Fig. 2.4d–e). Spectroscopic and computational results indicate that Cu(II) carboxylate coordination species on the catalyst surface can effectively stabilize active $Cu^{\delta+}$ species [44]. Additionally, heteroatom doping can also stabilize $Cu^{\delta+}$ species. For example, the average oxidation state of copper can range from + 0.25 to + 0.78 by varying the boron doping content [45]. PDOS analysis shows that electron transfer from Cu to boron creates a negatively charged Cu site, increasing the formation energy of *CHO while lowering the reaction barrier for *CO dimerization (Fig. 2.4f). These findings provide new insights into enhancing the stability of active $Cu^{\delta+}$ species during ECR.

However, in situ reduction of OD-Cu can also cause morphological and structural changes, making the identification of its true active sites a subject of ongoing debate [46–48]. For instance, Cu^+, mixed oxidation states ($Cu^{\delta+}$) of Cu (e.g., Cu^{2+} and Cu^+) or their interface have been identified as the active sites for promoting CO–CO coupling in OD-Cu during ECR [41, 49]. Conversely, another study suggests that metallic Cu is the active site in bulk Cu catalysts, attributing this to the instability of subsurface oxides under negative cathodic potentials [50]. Additionally, the active site of Cu nanoparticles (NPs) is reported to be metallic Cu nanograin, as investigated through the structural dynamics during the life cycle of Cu NPs using liquid cell scanning transmission electron microscopy [51]. Therefore, the real active sites and structural evolution of OD-Cu under ECR reaction conditions remain unclear.

2.2.3 Grain Boundaries

Grain boundaries (GBs), as structural defects, can create significant structural perturbations in their vicinity, inducing dislocations in the GBs region and leading to lattice strains (either compressive or tensile strains). In other words, GBs can generate high-energy surfaces through stabilized dislocations, which are thermodynamically favorable for catalysis [52]. To enhance the selectivity for multi-carbon products, various strategies have been developed to engineer GBs structures. These structures can adjust the adsorption energy of key intermediates on the catalyst surface, thus enhancing product selectivity [53, 54]. For example, Chen et al. used PVP as an additive to control the growth rate of Cu during electrodeposition, resulting in Cu with abundant GBs (Fig. 2.5a). This catalyst achieves 70% selectivity for ethylene and ethanol across a wide potential range [23]. Han and colleagues found that the oxide/hydroxide crystals in HQ-Cu (containing Cu, Cu_2O, CuO) and AN-Cu (containing Cu, $Cu(OH)_2$) fragment into nanosized irregular Cu grains under applied cathodic potentials (Fig. 2.5b). This fragmentation results from oxidation–reduction cycling, which not only creates a complex GBs network but also exposes various

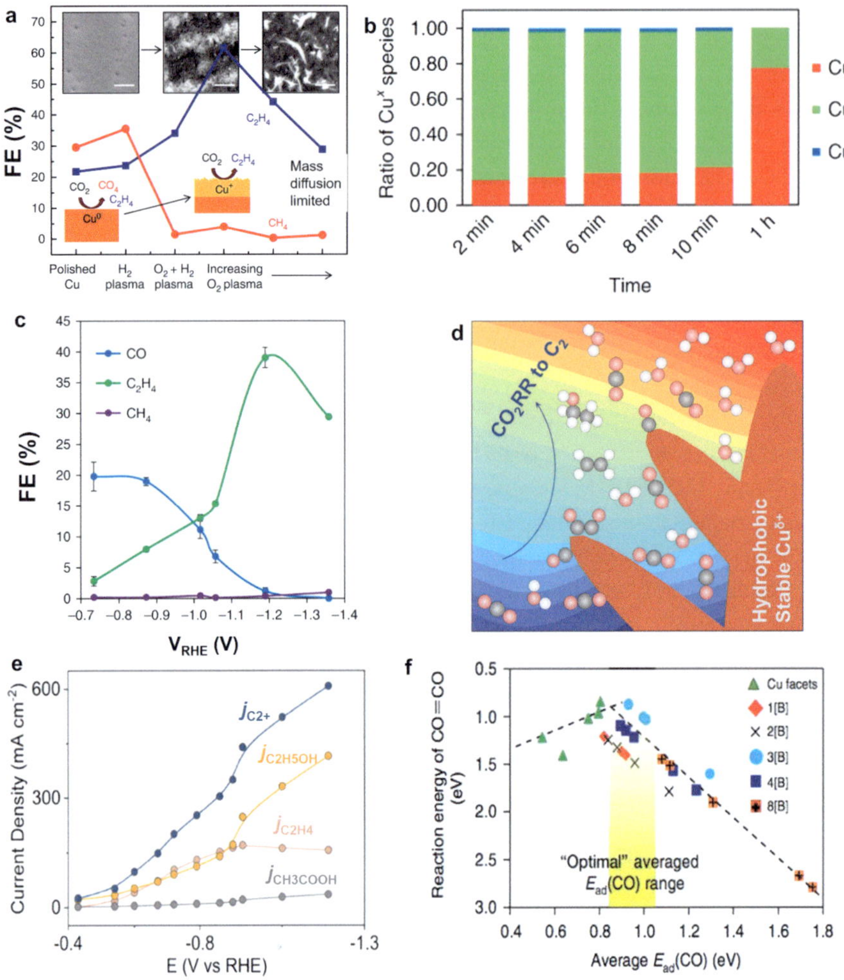

Fig. 2.4 Oxidation states manipulation for ECR. **a** Summary of the C_2H_4 selectivity of different plasma-activated Cu electrocatalysts [41]. Reprinted with permission. Copyright 2016 Springer Nature. **b** Ratio of Cu oxidation states relative to the reaction time at − 1.2 V versus RHE for electro-redeposited Cu [42]. Reprinted with permission. Copyright 2018 Springer Nature. **c** Reduced products of ECR on electro-redeposited Cu [42]. Reprinted with permission. Copyright 2018 Springer Nature. **d** Illustration of the ECR process on the ultrastable and hydrophobic $Cu^{\delta+}$ catalyst [44]. Reprinted with permission. Copyright 2023 American Chemical Society. **e** Current density distribution of the ultrastable and hydrophobic $Cu^{\delta+}$ catalyst [44]. Reprinted with permission. Copyright 2023 American Chemical Society. **f** Boron doping was able to change the average absorption energy of CO in the copper surface and lower the CO–CO dimerization energy [45]. Reprinted with permission. Copyright 2018 Springer Nature

high-index facets. These structural features facilitate C–C coupling, thereby enhancing the selectivity for C_{2+} products [21]. Huang et al. prepared La_2CuO_4 nanobamboo perovskites with abundant twin boundaries, achieving a 60% Faradaic efficiency (FE) for ethylene. The high ethylene selectivity is attributed to the active (113) surface with inherent strain effects (Fig. 2.5c) [24]. Therefore, designing and constructing unique GBs structures is highly beneficial for ECR reactions.

In complex GBs systems, single strategy often fails to meet practical needs in catalyst design. Combining multiple structural defects as active sites is often necessary to construct complex GBs structures, where various structural characteristics synergistically enhance the activity and stability of catalysts [42, 56, 57]. For OD-Cu, the active Cu^+ species inevitably undergoes electrochemical reduction to bulk metallic Cu at high current densities [58], leading to poor selectivity for C_{2+} products. This presents a significant challenge for the effective utilization of OD-Cu. However, research indicates that abundant nanograin boundaries can enhance the stability of catalyst morphology and Cu^0/Cu^+ interfaces under high polarization and high current densities, preventing catalyst reconstruction and improving catalytic stability (Fig. 2.5d) [55]. Moreover, nanograin boundaries and Cu^0/Cu^+ interfaces can increase *CO adsorption strength, promoting CO–CO coupling toward C_{2+} products [51, 59]. Therefore, constructing intricate structures assembles with abundant nanograin boundaries and Cu^0/Cu^+ interfaces in Cu-based catalysts holds promise for achieving efficient ECR to C_{2+} products. Additionally, oxygen vacancies, a well-known defect, are widely employed to tailor the properties of catalysts [60–62]. For example, due to their weakly bounded electrons, oxygen vacancies serve as excellent Lewis base sites to enhance the binding affinities of key intermediates (such as *CO and *COH), thereby promoting the production of C_2 (Fig. 2.5e) [56]. Moreover, a higher concentration of oxygen vacancies near active sites is favorable for activating CO_2 molecules, lowering the reaction barrier for target products [63]. Therefore, integrating oxygen vacancies with grain boundaries and interface engineering is likely to modulate the surface electronic structure and concentration of active sites, thus accelerating reaction kinetics. However, effectively constructing oxygen vacancies and interface structures on catalyst surface remains highly challenging. Gaining a deeper understanding of the synergistic interactions between oxygen vacancies and interface structures will offer valuable guidance for the rational design of catalysts.

2.2.4 Morphology Modification

The morphology of Cu nanostructures, such as size, shape, and crystal facets, significantly impacts the selectivity of ECR towards multi-carbon products [25, 64, 65]. The Cu(100) facet favors ethylene production, whereas the Cu(111) facet promotes methane production [65]. Buonsanti et al. synthesized Cu nanocrystals with varying sizes and shapes using a consistent colloidal method. The results demonstrated that cubic Cu nanocrystals exhibit

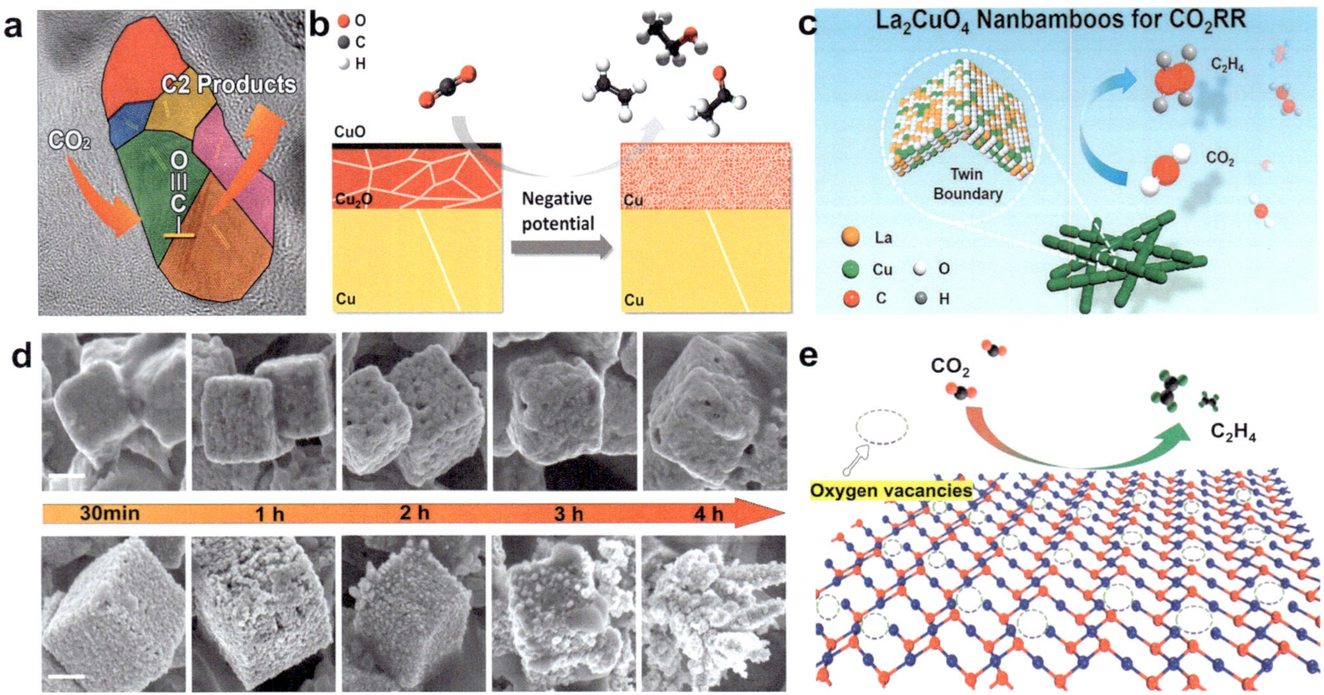

Fig. 2.5 Grain boundaries for ECR. **a** Illustration of the ECR process on the grain boundaries-rich Cu catalyst [23]. Reprinted with permission. Copyright 2020 American Chemical Society. **b** The fragmentation process of oxide/hydroxide crystals into nanosized irregular Cu grains under the applied negative potentials [21]. Reprinted with permission. Copyright 2020 American Chemical Society. **c** The La_2CuO_4 nanobamboo (La_2CuO_4 NBs) perovskite with rich twin boundaries showing a high FE of 60% toward ethylene [24]. Reprinted with permission. Copyright 2021 American Chemical Society. **d** SEM images of dynamic structure evolution for $Cu_2O(CO)$ and Cu_2O electrocatalysts during ECR [55]. Reprinted with permission. Copyright 2023 American Chemical Society. **e** Schematic of electrocatalytic reduction of CO_2 on Vo-rich CuO_x–Vo surface to C_2H_4 [56]. Reprinted with permission. Copyright 2018 WILEY–VCH Verlag GmbH & Co. KGaA, Weinheim

higher ethylene selectivity than spherical ones, primarily due to the dominant presence of the Cu(100) facet [25]. Among the cubic Cu nanocrystals, those with a 44 nm edge length exhibit the highest ECR activity and ethylene selectivity compared to cubes with edge lengths of 24 nm and 63 nm, revealing a non-monotonic size-dependent selectivity in cubic Cu nanocrystals (Fig. 2.6a) [25]. Mild plasma treatment introduces oxygen/chlorine ions into Cu nanocubes, the surface and subsurface oxygen species could influence the binding strength of CO on the surface, promoting the formation of multi-carbon products such as ethylene, ethanol, and isopropanol [66]. Song et al. reported that branched CuO nanoparticles achieve over 70% ethylene selectivity after activation. Compared to the cubic morphology, the initial branched CuO structure forms highly active interfaces and junctions in-between during activation, resulting in a large surface area and high local pH (Fig. 2.6b) [67]. Yang and colleagues demonstrated that densely packed Cu NPs, after undergoing structural rearrangement during the ECR reaction, transform into a mixture of highly active cubic particles and smaller nanoparticles. At lower overpotentials (− 0.53 V vs. RHE), selective formation of C_2–C_3 products can be achieved [68].

Additionally, when high-curvature structures such as nanoneedles and nanodendrites are involved, they significantly impacts the catalyst's structure and catalytic mechanism [70]. For instance, nanoneedles can promote the nucleation of smaller gas bubble [71], facilitating field-induced reactant concentration [72, 73]. The high local electric field can cause the accumulation of charged cations, stabilizing reaction intermediates and enhancing ECR activity [74]. Edward et al. reported that the morphology of Cu derived from a sol–gel is controlled by electro-redeposition. Depending on the applied potentials, different structural morphologies, including nanoclusters, nanoneedles, nanowhiskers, and nanodendrites, appear sequentially (Fig. 2.6c). This growth results from the simultaneous dissolution and redeposition process of Cu ions from the bulk phase [42]. Inspired by gas-trapping phenomenon observed in spiders, hierarchically structured Cu dendrites undergo surface hydrophobic treatment with 1-octadecanethiol. The treated electrodes can capture CO_2 gas at the electrolyte–electrode interface, forming a triple-phase boundary (Fig. 2.6d), which effectively suppresses HER and promotes ECR reactions [69]. In summary, the morphology effect offers a promising paradigm for the design and development of Cu-based electrocatalysts.

2.3 The CO_2 Electrolyzer

There are three main configurations for CO_2 electrolyzers: H-cells, flow cell, and membrane electrode assembly (MEA).

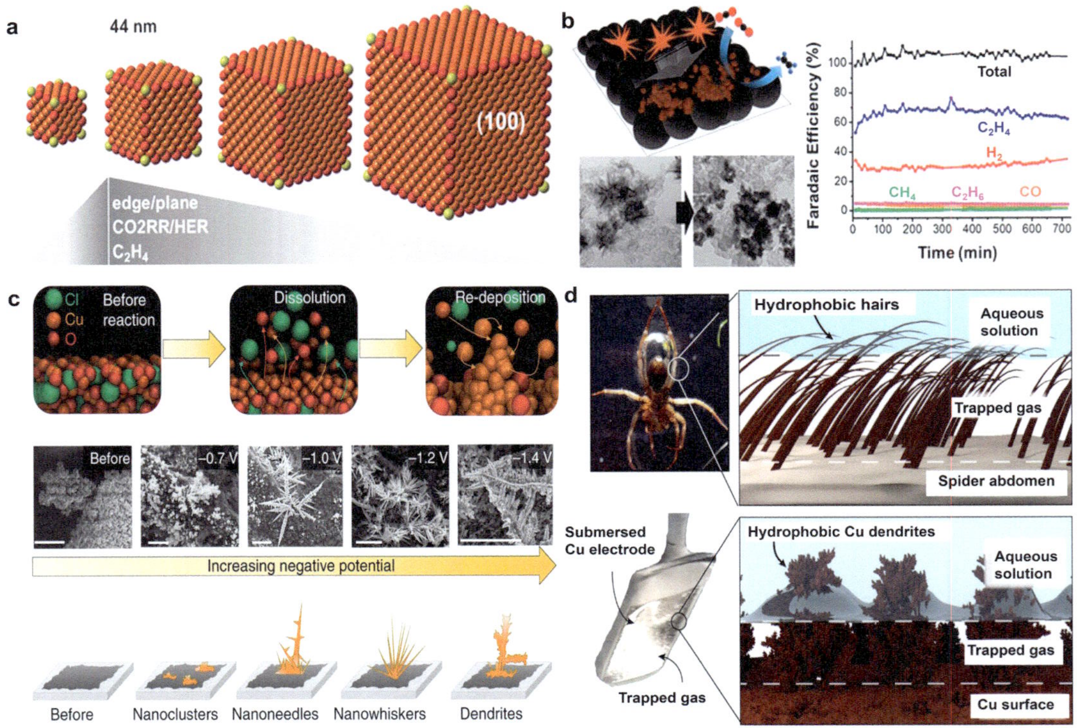

Fig. 2.6 Morphology effect for ECR. **a** Schematic of a non-monotonic size-dependence of the selectivity in cube-shaped copper nanocrystals [25]. Reprinted with permission. Copyright 2016 WILEY–VCH Verlag GmbH & Co. KGaA, Weinheim. **b** Schematic illustrating the transformation process of branched CuO NPs on conductive carbon materials during activation [67]. Reprinted with permission. Copyright 2019 American Chemical Society. **c** Schematic of the electro-growth process, whereby simultaneous dissolution and redeposition of Cu results in structured deposits [42]. Reprinted with permission. Copyright 2018 Springer Nature. **d** The plastron effect: the use of a hydrophobic surface to trap a layer of gas between the solution–solid interface. This is illustrated on a diving bell spider for subaquatic breathing and on a hydrophobic dendritic Cu surface for aqueous CO_2 reduction [69]. Reprinted with permission. Copyright 2019 Springer Nature

2.3.1 H-cells

Most initial performance evaluations of ECR and catalyst screening are typically conducted using an H-cell. This system involves three electrodes: the working electrode and reference electrode are placed in the cathode chamber, while the counter electrode is fixed in the anode chamber. The cathode and anode chambers are separated by an ion-exchange membrane (Fig. 2.7) [75, 76]. This simple setup facilitates rapid catalyst screening. CO_2 is bubbled into the cathode chamber to maintain a CO_2-saturated electrolyte. Although some ECR catalysts have shown excellent performance in H-cells (Table 2.1), their practical application is constrained by several factors. The CO_2 concentration near the catalyst is limited by its solubility (approximately 34 mM at room temperature), resulting in mass transport constraints, which cap the maximum operating current at below 100 mA cm^{-2} [77]. ECR performance testing is typically conducted under neutral conditions to avoid side reactions, such as carbonate formation in alkaline electrolytes and HER in acidic electrolytes [76]. However, the ECR reaction and pathways are highly dependent on the local reaction microenvironment, which influences the reaction energetics on the catalyst surface [75, 78]. Consequently, catalysts developed in H-cells may perform significantly differently under industrial conditions due to variations in reaction microenvironments. Furthermore, impurities deposition on the catalyst surface can affect the ECR reaction pathways, leading to catalyst deactivation [79, 80]. Therefore, alternative testing systems are needed to evaluate the catalyst performance under industrially relevant ECR conditions.

2.3.2 Flow Cells

To address the challenges of low CO_2 solubility, low current density, and mass transport limitations in H-cells, gas diffusion electrodes (GDEs) have been developed. GDEs significantly shorten the diffusion path of CO_2 to the electrode surface (approximately 50 nm), enhancing mass transport efficiency and enabling the industrial application of ECR [78, 81]. Unlike H-cells, flow cells can effectively utilize alkaline electrolytes, which are unsuitable for H-cells. GDEs allow separation of the CO_2 flow field from the liquid electrolytes, maintaining a high local pH on the catalyst surface, which inhibits HER and promotes C–C coupling [82–84]. This configuration allows for high product selectivity at high current densities, positioning flow cells as a potential replacement for H-cells in ECR performance testing. In the flow cell configuration (Fig. 2.8a), the catholyte and anolyte circulate through the cathode and anode chambers, respectively. The cathode and anode chambers are separated by an ion exchange membrane. Gaseous CO_2 is continuously supplied to the cathode catalyst [85, 86].

Current research on ECR catalysts has transitioned from H-cells to flow cells. The complexity of flow cell configurations has driven the evolution of catalyst material design

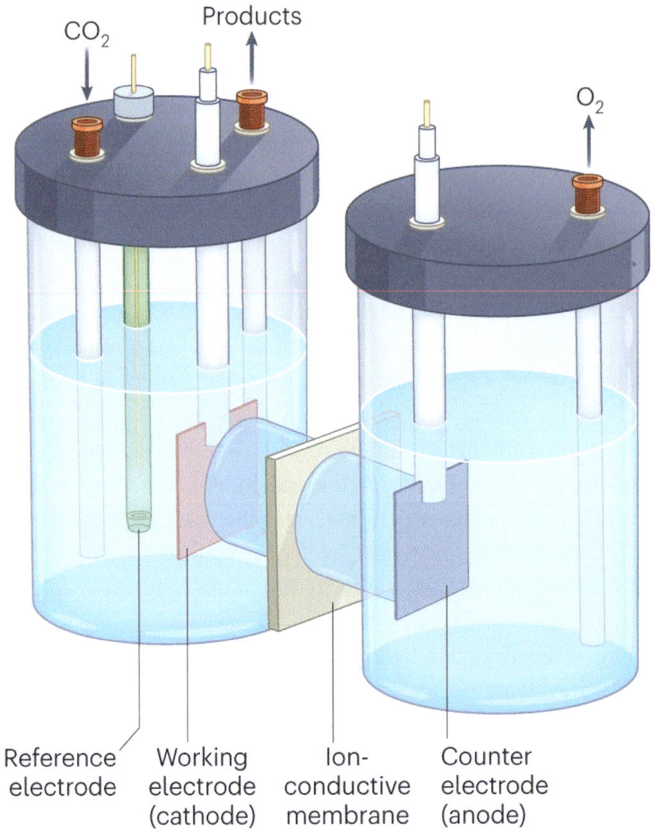

Fig. 2.7 Schematic illustration of three-electrode H-cells [76]. Reprinted with permission. Copyright 2023 Springer Nature

into a multifaceted approach, encompassing the catalyst, the local microenvironment, and the overall system (Table 2.2). The catholyte can regulate the local microenvironment on the catalyst surface, optimizing CO_2 conversion efficiency (Fig. 2.8b). For example, Dinh et al. showed that a high concentration of KOH electrolyte can significantly reduce the onset potential for ECR [88]. Chen and colleagues reported that Cu-polyamine in a strong alkaline electrolyte (10 M KOH) can achieve an 87% FE for ethylene [82]. Elevated pH levels can reduce H coverage on the catalyst surface, thus inhibiting HER. This enhanced catalytic activity can, in turn, lower the cell voltage, thereby improving energy efficiency (Fig. 2.8c). Although strong alkaline electrolytes can accelerate the formation of multi-carbon products [88], they also lead to low CO_2 utilization efficiency and irreversible electrolyte acidification [89, 90], creating challenges for optimal performance. In acidic electrolytes, the FE of HER side reaction is typically 20% or higher, making it energy-intensive and inhibitory to the ECR reaction [91]. To address these issues, neutral

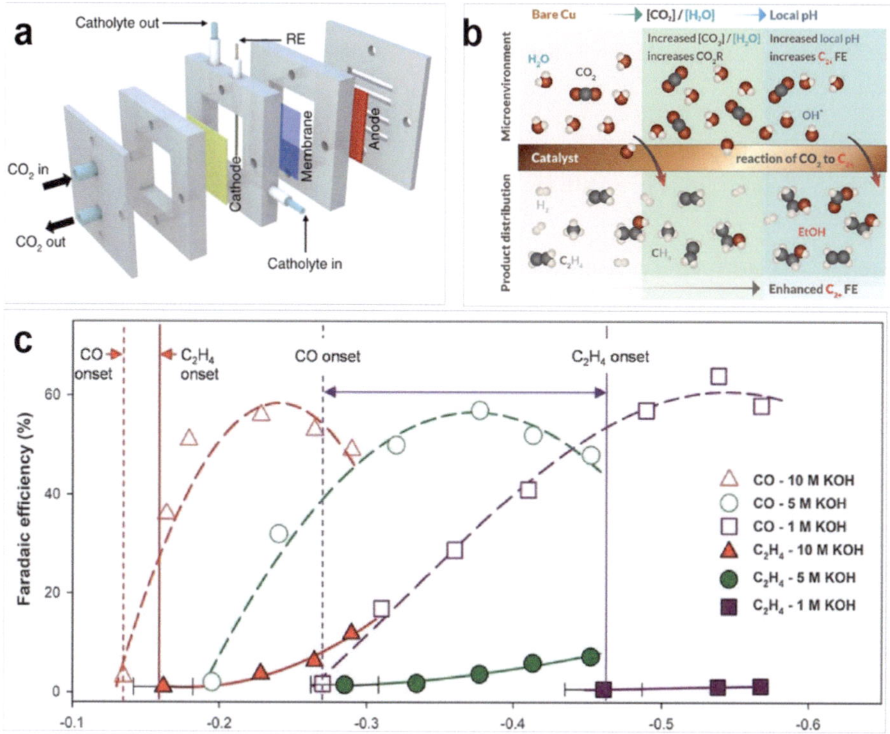

Fig. 2.8 **a** Schematic illustration of flow cells [86]. Reprinted with permission. Copyright 2019 Springer Nature. **b** Schematic illustration of the local microenvironment and its effect on ECR selectivity [87]. Reprinted with permission. Copyright 2022 American Chemical Society. **c** C_2H_4 and CO FEs showing the reduction of C_2H_4 onset potential with increasing KOH concentrations [88]. Reprinted with permission. Copyright 2018 The American Association for the Advancement of Science

electrolytes like potassium bicarbonate have been employed, though product selectivity of ECR in neutral electrolyte is lower than in alkaline environments [92, 93]. Overall, the flow cell configurations overcome the limitations of H-cells, facilitating the large-scale application of ECR.

2.3.3 Membrane Electrode Assemblies

In a flow cell, liquid electrolytes in the cathode and anode chambers create high resistance, lowering the overall energy efficiency of the system. To tackle this issue, a practical approach is to remove the liquid electrolyte between the electrodes, leading to the development of the membrane electrode assembly (MEA) [118]. An MEA operates as a

Table 2.2 The reported performance of ECR in different CO_2 electrolyzer

Electrolyzer	Catalyst	Electrolyte	V (vs. RHE/ full-cell voltage)	Current density (mA cm^{-2})	Stability (h)	Main product (FE)	References
H-cell	Cu NPs	0.1 M $KHCO_3$	1.7 versus Ag/AgCl	7.5	/	C_2H_4 (92.8%)	[94]
	CuGNC-VL	0.5 M $KHCO_3$	− 0.87 V_{RHE}	10.4	/	Ethanol (70.5%)	[95]
	$K_{11.2\%}$-Cu_2Se nanosheets	0.1 M $KHCO_3$	− 0.8 V_{RHE}	~50	130	Ethanol (70%)	[96]
	Carbon-supported Cu	0.1 M $KHCO_3$	− 0.6 V_{RHE}	1.5	16	Ethanol (91%)	[97]
	GN/ZnO/Cu_2O	0.5 M $NaHCO_3$	− 0.9 V versus Ag/AgCl	/	/	n-propanol (30%)	[98]
	Cu nanoparticles	0.1 M $KHCO_3$	− 0.81 V_{RHE}	12.8	/	n-propanol (5.9%)	[68]
	Nitrogen-doped nanodiamond (NDD)	0.5 M $NaHCO_3$	− 0.8 to − 0.1 V_{RHE}	~7	3	Acetate (77%)	[99]
	Sn coupled with defective CuO	[Bmim]BF_4/ H_2O	− 2.0 V versus Ag/Ag^+	67	36	Methanol (88.6%)	[100]
	Pulse electrodeposited Cu	0.5 M $NaHCO_3$	− 2.8 V_{SCE}	38	/	CH_4 (85%)	[101]
	3D core–shell porous-structured Cu@Sn	0.5 M $KHCO_3$	− 0.93 V_{RHE}	16.52	15	Formate (100%)	[102]
Flow cell	Cu-polyamine	10 M KOH	0.47 V_{RHE}, 2.02 full cell	/	/	C_2H_4 (87%)	[82]
	Cu-Al	1 M KOH	1.5 V_{RHE}	400	50	C_2H_4 (80%)	[103]
	Graphite/carbon NPs/Cu	7 M KOH	0.55 V_{RHE}	55–70	150	C_2H_4 (70%)	[88]

(continued)

Table 2.2 (continued)

Electrolyzer	Catalyst	Electrolyte	V (vs. RHE/ full-cell voltage)	Current density (mA cm^{-2})	Stability (h)	Main product (FE)	References
	CuO nanoplate	0.5 M KCl	~5.0	150	55	C_2H_4 (74%)	[104]
	Antiswelling AEI-modified oxide-derived Cu nanosheets	1 M KOH	− 0.8 V_{RHE}	800	9	C_2H_4 (65%)	[105]
	Cu-PFSA	7 M KOH	− 0.91 V_{RHE}	1550	/	C_2H_4 (60%)	[85]
	Defect-side-rich Cu	1 M KOH	− 0.95 V_{RHE}	100	/	Ethanol (52%)	[106]
	Hg-CoTPP with N-doped graphene	1 M KOH	− 0.8 V_{RHE}	420	360	CO (~100%)	[107]
	Cu NPs	1 M $KHCO_3$	− 0.97 V_{RHE}	300	/	n-propanol (4.6%)	[108]
	Grain boundary-enriched bismuth	1 M $KHCO_3$	− 0.77 V_{RHE}	450	/	Formate (97%)	[109]
	$CuGa_2$	1 M KOH	− 0.3 V_{RHE}	21.4	24	Methanol (78%)	[110]
	Copper(I)-based coordination polymer	1 M KOH	− 0.9 V_{RHE}	350	5	CH_4 (75%)	[111]
	$CuGaO_2$ nanosheet	1 M KOH	− 2 V_{RHE}	1000	5	CH_4 (68–55%)	[112]
MEA	CuO nanoplate	0.5 M $KHCO_3$	− 3.2	200	/	C_2H_4 (80%)	[104]
	Cu-SiOx	0.1 M $KHCO_3$	− 4.1	215	50	C_2H_4 (65%)	[113]
	Cu_3Sn	1 M KOH	− 3	900	48	Ethanol (40%)	[114]
	Copper(II) phthalocyanine	0.05 M $KHCO_3$	− 4.2	190	110	CH_4 (56%)	[115]

(continued)

Table 2.2 (continued)

Electrolyzer	Catalyst	Electrolyte	V (vs. RHE/ full-cell voltage)	Current density (mA cm^{-2})	Stability (h)	Main product (FE)	References
	Cu nanoparticles supported on N-doped carbon	0.1 M $KHCO_3$	− 4	230	50	CH_4 (60%)	[116]
	Ni single atom supported on porous carbon	0.5 M $KHCO_3$	− 2.6	140	9	CO (99%)	[117]

dual-electrode zero-gap cell, with a polymer electrolyte membrane sandwiched between the anode and cathode. The electrolyte circulates through the porous anode without the need for a reference electrode. Humidified CO_2 is continuously supplied to the cathode's GDE (Fig. 2.9a). Compared to flow cells, the MEA configuration leverages the gas diffusion electrode (GDE) of a flow cell, minimizing the electrolyte usage by enabling direct contact between the catalyst layer and the ion exchange membrane. This design mitigates cell resistance, boosts energy efficiency, and enhances cell stability, making it more feasible for practical applications [90].

The formation of multi-carbon products in an MEA electrolyzer has yet to be extensively studied (Table 2.1). Gabardo et al. demonstrate the potential for multi-carbon product formation in an MEA reactor by sputtering Cu onto a porous PTFE membrane as the catalyst and adjusting the reaction temperature to 40 °C, achieving 30% ethylene selectivity [119]. Gong et al. reported that oxygen-rich ultrathin CuO nanoplate arrays could achieve 80% ethylene selectivity at a current density of 200 mA cm^{-2} (Fig. 2.9b). Compact evaporated Cu films and stable nanostructures effectively prevent GDL flooding during ECR testing, enhancing the long-term stability of the catalyst [104]. Zheng et al. reported that a low-entropy Cu_3Sn catalyst maintains stability for over 48 h at industrial-scale current densities, with 40% ethanol selectivity (Fig. 2.9c) [114]. Gu et al. constructed a defect-rich surface by electrodepositing Cu in a CO-rich atmosphere, significantly promoting CO_2 conversion to ethanol, achieving 70% FE for C_{2+} alcohols at a partial current density of 100 mA cm^{-2}. In a 30-h stability test, the MEA electrolyzer produces 500 mg of C_{2+} alcohols per cm^2 of catalyst (Fig. 2.9d) [106]. In summary, MEA electrolyzers may offer better stability than flow cells and H-cells, making them more attractive for commercial applications.

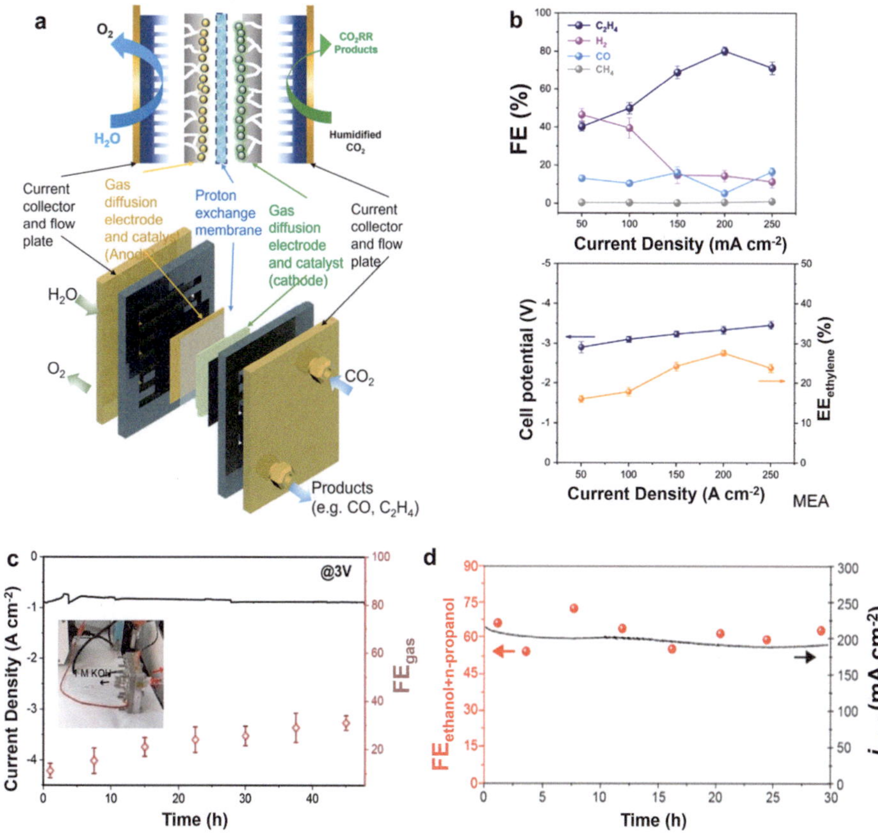

Fig. 2.9 **a** Schematic of membrane-electrode assembly for ECR [118]. Reprinted with permission. Copyright 2022 Elsevier. **b** Gas product FEs, cell potential and energy efficiency for C_2H_4 of the DVL-Cu@GDL catalyst in an MEA electrolyzer [104]. Reprinted with permission. Copyright 2022 Springer Nature. **c** The long-term stability test of Cu_3Sn in 1 M KOH at 3 V [114]. Reprinted with permission. Copyright 2021 WILEY–VCH Verlag GmbH & Co. KGaA, Weinheim. **d** ECR to ethanol using Cu-DS catalyst in a MEA system during 30 h of electrolysis under a full-cell voltage of 3.5 V [106]. Reprinted with permission. Copyright 2021 Elsevier

2.4 Gas Diffusion Electrodes

In an H-cell, the catalyst is typically deposited on a substrate such as carbon paper, glassy carbon rods, or directly on a metal foil as the working electrode. In aqueous electrolytes, this setup leads to sluggish mass transfer and low ECR reaction rates [120, 121]. To enhance ECR performance, GDEs have been introduced, enabling CO_2 to diffuse directly to the electrode–electrolyte interface [122]. GDEs typically consist of a gas diffusion layer

(GDL) and a catalyst layer (Fig. 2.10a) [88, 123]. This structure forms a three-phase interface where the liquid electrolyte and gaseous CO_2 coexist in the catalyst layer [122]. The GDL contains a dense carbonaceous network that serves as a porous support for the catalyst layer. This dense network, typically made from carbon fibers like carbon paper or PTFE, ensures mechanical stability and facilitates efficient gas transport. The dense carbon network in the GDL plays a crucial role in forming a stable and efficient interface for the electrochemical reactions in ECR systems [124]. The GDLs are typically treated with PTFE to enhance hydrophobicity, effectively preventing water infiltration. The nanoporous structure of the GDL ensures the transport of CO_2 gas to the catalyst layer. Maintaining GDE hydrophobicity is crucial during the ECR reaction [125, 126]. Loss of hydrophobicity can cause electrolyte permeation into GDL pores, leading to flooding within the GDL. This obstructs CO_2 diffusion, inhibiting ECR and promoting HER [127, 128]. Therefore, GDE stability is critical for long-term stability testing of CO_2 electrolyzers.

To overcome these challenges, researchers have incorporated hydrophobic PTFE frameworks into GDEs. Compared to carbon paper-based GDEs, hydrophobic PTFE exhibits better long-term stability during ECR testing. Sargent et al. sputtered Cu NPs onto porous PTFE and coated the catalyst surface with carbon nanoparticles and graphene to enhance conductivity, thereby creating a stable electrode interface (Fig. 2.10b). This setup achieved 70% ethylene selectivity at current densities of 75 to 100 mA cm^{-2} in 7 M KOH electrolyte, with extended long-term stability testing up to 150 h without decay—300 times longer than traditional GDEs (Fig. 2.10c) [88]. This GDE design has demonstrated structural superiority. Additionally, Han et al. coupled GDEs with 3D nanostructured catalysts to design a 3D Cu-chitosan (CS)-GDL electrode. CS acts as a transition layer between the Cu catalyst and the GDL (Fig. 2.10d). The highly interlaced network promotes the growth of 3D Cu films, facilitating rapid electron transfer and reducing mass transfer limitations. This structure achieved 88.2% C_{2+} selectivity at a current density of 900 mA cm^{-2} (Fig. 2.10e) [129]. These advancements in GDE design underscore the importance of creating a stable and efficient interface for electrochemical reactions, highlighting the critical role of dense carbonaceous networks and hydrophobic treatments in achieving high performance and long-term stability in ECR systems.

In summary, ECR reactions depend on both catalyst design and reactor configuration, both of which can significantly influence the reaction products. Advancing ECR technology requires clarifying the true active sites of the catalyst within the reactor and understanding their relationship with ECR selectivity. Furthermore, understanding the dynamic evolution of active sites under actual reaction conditions is crucial. This approach will facilitate the rational design of catalysts and optimize reactor design, thereby accelerating the commercialization of ECR technology.

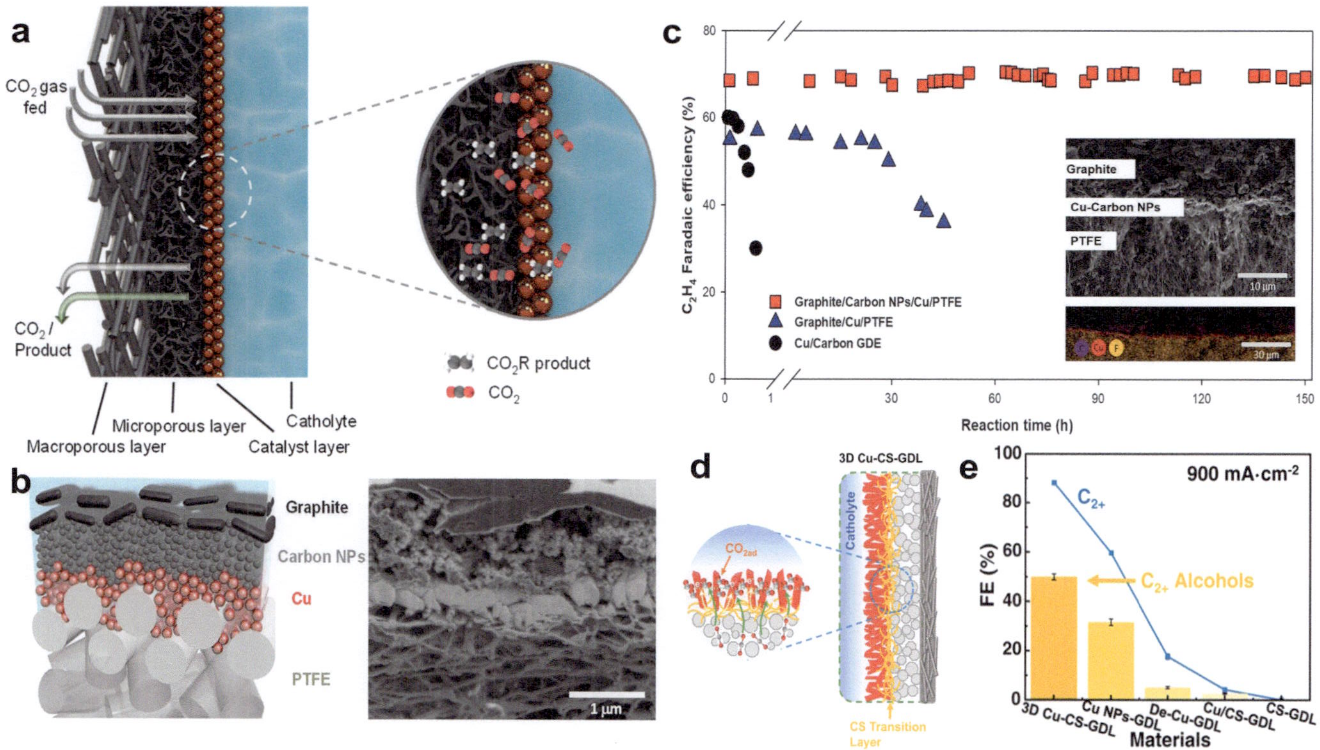

Fig. 2.10 **a** Schematic of an ECR GDE cathode and its fundamental components [8]. Reprinted with permission. Copyright 2024 American Chemical Society. **b** Structure of the polymer-based gas diffusion electrode [88]. Reprinted with permission. Copyright 2018 The American Association for the Advancement of Science. **c** The long-term stability test of polymer-based gas diffusion electrode [88]. Reprinted with permission. Copyright 2018 The American Association for the Advancement of Science. **d** Schematic of 3D Cu–CS-GDL electrode [129]. Reprinted with permission. Copyright 2023 Springer Nature. **e** The FE of C_{2+} products and C_{2+} alcohols on different GDEs at a current density of 900 mA cm^{-2} [129]. Reprinted with permission. Copyright 2023 Springer Nature

Competing Interests The authors have no conflicts of interests to declare that are relevant to the content of this chapter.

References

1. Yoro, K.O., Daramola, M.O.: CO_2 emission sources, greenhouse gases, and the global warming effect. In: Advances in Carbon Capture, pp. 3–28. Elsevier (2020)
2. Chu, S., et al.: The path towards sustainable energy. Nat. Mater. **16**(1), 16–22 (2017)
3. Donner, S., Zickfield, K.: Canada's contribution to meeting the temperature limits in the Paris climate agreement (2016)
4. Environment; Canada, C.C.: National inventory report 1990–2018: greenhouse gas sources and sinks in Canada. Environment Canada Ottawa (2018)
5. Bushuyev, O.S., et al.: What should we make with CO_2 and how can we make it? Joule. **2**(5), 825–832 (2018)
6. Lu, Q., Jiao, F.J.N.E.: Electrochemical CO_2 reduction: electrocatalyst, reaction mechanism, and process engineering. **29**, 439–456 (2016)
7. Benson, E.E., et al.: Electrocatalytic and homogeneous approaches to conversion of CO_2 to liquid fuels. Chem. Soc. Rev. **38**(1), 89–99 (2009)
8. O'Brien, C.P., et al.: CO_2 electrolyzers. Chem. Rev. (2024)
9. Dau, H., et al.: The mechanism of water oxidation: from electrolysis via homogeneous to biological catalysis. ChemCatChem **2**(7), 724–761 (2010)
10. Kortlever, R., et al.: Catalysts and reaction pathways for the electrochemical reduction of carbon dioxide. J. Phys. Chem. Lett. **6**(20), 4073–4082 (2015)
11. Zhang, Y.-J., et al.: Competition between CO_2 reduction and H_2 evolution on transition-metal electrocatalysts. ACS Catal. **4**(10), 3742–3748 (2014)
12. Teeter, T.E., Van Rysselberghe, P.: Reduction of carbon dioxide on mercury cathodes. J. Chem. Phys. **22**(4), 759–760 (1954)
13. Hori, Y., et al.: Production of CO and CH_4 in electrochemical reduction of CO2 at metal electrodes in aqueous hydrogencarbonate solution. Chem. Lett. **14**(11), 1695–1698 (1985)
14. Costentin, C., et al.: Catalysis of the electrochemical reduction of carbon dioxide. Chem. Soc. Rev. **42**(6), 2423–2436 (2013)
15. Appel, A.M., et al.: Frontiers, opportunities, and challenges in biochemical and chemical catalysis of CO_2 fixation. Chem. Rev. **113**(8), 6621–6658 (2013)
16. Bagger, A., et al.: Electrochemical CO_2 reduction: a classification problem. ChemPhysChem **18**(22), 3266–3273 (2017)
17. Hori, Y.I.: Electrochemical CO_2 reduction on metal electrodes. In: Modern Aspects of Electrochemistry, pp. 89–189 (2008)
18. Kuhl, K.P., et al.: Electrocatalytic conversion of carbon dioxide to methane and methanol on transition metal surfaces. J. Am. Chem. Soc. **136**(40), 14107–14113 (2014)
19. Hoang, T.T., et al.: Nanoporous copper–silver alloys by additive-controlled electrodeposition for the selective electroreduction of CO_2 to ethylene and ethanol. J. Am. Chem. Soc. **140**(17), 5791–5797 (2018)
20. Chang, Z., et al.: The tunable and highly selective reduction products on Ag@Cu bimetallic catalysts toward CO_2 electrochemical reduction reaction. J. Phys. Chem. C **121**(21), 11368–11379 (2017)
21. Lei, Q., et al.: Investigating the origin of enhanced C_{2+} selectivity in oxide-/hydroxide-derived copper electrodes during CO_2 electroreduction. J. Am. Chem. Soc. **142**(9), 4213–4222 (2020)

22. Lee, S.Y., et al.: Mixed copper states in anodized Cu electrocatalyst for stable and selective ethylene production from CO_2 reduction. J. Am. Chem. Soc. **140**(28), 8681–8689 (2018)
23. Chen, Z., et al.: Grain-boundary-rich copper for efficient solar-driven electrochemical CO_2 reduction to ethylene and ethanol. J. Am. Chem. Soc. **142**(15), 6878–6883 (2020)
24. Wang, J., et al.: Grain-boundary-engineered La_2CuO_4 perovskite nanobamboos for efficient CO_2 reduction reaction. Nano Lett. **21**(2), 980–987 (2021)
25. Loiudice, A., et al.: Tailoring copper nanocrystals towards C_2 products in electrochemical CO_2 reduction. Angew. Chem. Int. Ed. **55**(19), 5789–5792 (2016)
26. Ma, M., et al.: Controllable hydrocarbon formation from the electrochemical reduction of CO_2 over Cu nanowire arrays. Angew. Chem. Int. Ed. **55**(23), 6680–6684 (2016)
27. Xue, D., et al.: Defect engineering on carbon-based catalysts for electrocatalytic CO_2 reduction. Nano-Micro Lett. **13**(1), 1–23 (2021)
28. Moura de Salles Pupo, M., Kortlever, R.: Electrolyte effects on the electrochemical reduction of CO_2. ChemPhysChem. **20**(22), 2926–2935 (2019)
29. Jia, Y., et al.: Cu-based bimetallic electrocatalysts for CO_2 reduction. Adv. Powder Mater. **1**(1), 100012 (2022)
30. Zhu, W., et al.: Formation of enriched vacancies for enhanced CO_2 electrocatalytic reduction over AuCu alloys. ACS Energy Lett. **3**(9), 2144–2149 (2018)
31. Zhang, X., et al.: Formation of lattice-dislocated bismuth nanowires on copper foam for enhanced electrocatalytic CO_2 reduction at low overpotential. Energy Environ. Sci. **12**(4), 1334–1340 (2019)
32. Zhu, W., et al.: Bimetallic electrocatalysts for CO_2 reduction. Electrocatalysis., 105–125 (2020)
33. Kuhl, K.P., et al.: New insights into the electrochemical reduction of carbon dioxide on metallic copper surfaces. Energy Environ. Sci. **5**(5), 7050–7059 (2012)
34. Zhang, H., et al.: Computational and experimental demonstrations of one-pot tandem catalysis for electrochemical carbon dioxide reduction to methane. Nat. Commun. **10**(1), 3340 (2019)
35. Chang, C.-J., et al.: Dynamic reoxidation/reduction-driven atomic interdiffusion for highly selective CO_2 reduction toward methane. J. Am. Chem. Soc. **142**(28), 12119–12132 (2020)
36. Dong, W.J., et al.: Tailoring electronic structure of bifunctional Cu/Ag layered electrocatalysts for selective CO_2 reduction to CO and CH_4. Nano Energy **78**, 105168 (2020)
37. Chen, C., et al.: Cu-Ag tandem catalysts for high-rate CO_2 electrolysis toward multicarbons. Joule. **4**(8), 1688–1699 (2020)
38. Lv, X., et al.: Electron-deficient Cu sites on Cu_3Ag_1 catalyst promoting CO_2 electroreduction to alcohols. Adv. Energy Mater. **10**(37), 2001987 (2020)
39. Ma, S., et al.: Electroreduction of carbon dioxide to hydrocarbons using bimetallic Cu–Pd catalysts with different mixing patterns. J. Am. Chem. Soc. **139**(1), 47–50 (2017)
40. Morales-Guio, C.G., et al.: Improved CO_2 reduction activity towards C_{2+} alcohols on a tandem gold on copper electrocatalyst. Nat. Catal. **1**(10), 764–771 (2018)
41. Mistry, H., et al.: Highly selective plasma-activated copper catalysts for carbon dioxide reduction to ethylene. Nat. Commun. **7**(1), 1–9 (2016)
42. De Luna, P., et al.: Catalyst electro-redeposition controls morphology and oxidation state for selective carbon dioxide reduction. Nat. Catal. **1**(2), 103–110 (2018)
43. Dattila, F., et al.: Active and selective ensembles in oxide-derived copper catalysts for CO_2 reduction. ACS Energy Lett. **5**(10), 3176–3184 (2020)
44. Fang, M., et al.: Hydrophobic, ultrastable $Cu^{\delta+}$ for robust CO_2 electroreduction to C_2 products at ampere-current levels. J. Am. Chem. Soc. **145**(20), 11323–11332 (2023)
45. Zhou, Y., et al.: Dopant-induced electron localization drives CO_2 reduction to C_2 hydrocarbons. Nat. Chem. **10**(9), 974–980 (2018)

46. Ren, D., et al.: Selective electrochemical reduction of carbon dioxide to ethylene and ethanol on copper (I) oxide catalysts. ACS Catal. **5**(5), 2814–2821 (2015)
47. Fan, L., et al.: 1D SnO_2 with wire-in-tube architectures for highly selective electrochemical reduction of CO_2 to C_1 products. Adv. Funct. Mater. **28**(17), 1706289 (2018)
48. Li, C.W., et al.: Electroreduction of carbon monoxide to liquid fuel on oxide-derived nanocrystalline copper. Nature **508**(7497), 504–507 (2014)
49. Liang, Z.-Q., et al.: Copper-on-nitride enhances the stable electrosynthesis of multi-carbon products from CO_2. Nat. Commun. **9**(1), 1–8 (2018)
50. Lum, Y., Ager, J.W.: Stability of residual oxides in oxide-derived copper catalysts for electrochemical CO_2 reduction investigated with 18O labeling. Angew. Chem. Int. Ed. **57**(2), 551–554 (2018)
51. Yang, Y., et al.: Operando studies reveal active Cu nanograins for CO_2 electroreduction. Nature **614**(7947), 262–269 (2023)
52. She, X., et al.: Grain-boundary surface terminations incorporating oxygen vacancies for selectively boosting CO_2 photoreduction activity. Nano Energy **84**, 105869 (2021)
53. Mariano, R.G., et al.: Selective increase in CO_2 electroreduction activity at grain-boundary surface terminations. Science **358**(6367), 1187–1192 (2017)
54. Pang, Y., et al.: Efficient electrocatalytic conversion of carbon monoxide to propanol using fragmented copper. Nat. Catal. **2**(3), 251–258 (2019)
55. Wu, Q., et al.: Nanograin-boundary-abundant Cu_2O-Cu nanocubes with high C_{2+} selectivity and good stability during electrochemical CO_2 reduction at a current density of 500 mA/cm^2. ACS Nano (2023)
56. Gu, Z., et al.: Oxygen vacancy tuning toward efficient electrocatalytic CO_2 reduction to C_2H_4. Small Methods. **3**(2), 1800449 (2019)
57. Wu, Z., et al.: Grain boundary and interface interaction Co-regulation promotes SnO_2 quantum dots for efficient CO_2 reduction. Chem. Eng. J. **451**, 138477 (2023)
58. Chen, J., Wang, L.: Effects of the catalyst dynamic changes and influence of the reaction environment on the performance of electrochemical CO_2 reduction. Adv. Mater. **34**(25), 2103900 (2022)
59. Yuan, X., et al.: Controllable Cu_0-Cu_+ sites for electrocatalytic reduction of carbon dioxide. Angew. Chem. **133**(28), 15472–15475 (2021)
60. Li, H., et al.: Oxygen vacancy structure associated photocatalytic water oxidation of BiOCl. ACS Catal. **6**(12), 8276–8285 (2016)
61. Li, H., et al.: Oxygen vacancy-mediated photocatalysis of BiOCl: reactivity, selectivity, and perspectives. Angew. Chem. Int. Ed. **57**(1), 122–138 (2018)
62. Li, X., et al.: Strategies for enhancing electrochemical CO_2 reduction to multi-carbon fuels on copper. Innov. Mater. (2023)
63. Wang, Y., et al.: Defect and interface engineering for aqueous electrocatalytic CO_2 reduction. Joule. **2**(12), 2551–2582 (2018)
64. Choi, C., et al.: A highly active star decahedron Cu nanocatalyst for hydrocarbon production at low overpotentials. Adv. Mater. **31**(6), 1805405 (2019)
65. Huang, Y., et al.: Electrochemical reduction of CO_2 using copper single-crystal surfaces: effects of CO* coverage on the selective formation of ethylene. ACS Catal. **7**(3), 1749–1756 (2017)
66. Gao, D., et al.: Plasma-activated copper nanocube catalysts for efficient carbon dioxide electroreduction to hydrocarbons and alcohols. ACS Nano **11**(5), 4825–4831 (2017)
67. Kim, J., et al.: Branched copper oxide nanoparticles induce highly selective ethylene production by electrochemical carbon dioxide reduction. J. Am. Chem. Soc. **141**(17), 6986–6994 (2019)

68. Kim, D., et al.: Copper nanoparticle ensembles for selective electroreduction of CO_2 to C_2–C_3 products. Proc. Natl. Acad. Sci. U.S.A. **114**(40), 10560–10565 (2017)
69. Wakerley, D., et al.: Bio-inspired hydrophobicity promotes CO_2 reduction on a Cu surface. Nat. Mater. **18**(11), 1222–1227 (2019)
70. Reller, C., et al.: Selective electroreduction of CO_2 toward ethylene on nano dendritic copper catalysts at high current density. Adv. Energy Mater. **7**(12), 1602114 (2017)
71. Thomas, B., et al.: Nanomorphology-enhanced gas-evolution intensifies CO_2 reduction electrochemistry (2017)
72. Liu, M., et al.: Enhanced electrocatalytic CO_2 reduction via field-induced reagent concentration. Nature **537**(7620), 382–386 (2016)
73. Saberi Safaei, T., et al.: High-density nanosharp microstructures enable efficient CO_2 electroreduction. Nano Lett. **16**(11), 7224–7228 (2016)
74. Chen, L.D., et al.: Electric field effects in electrochemical CO2 reduction. ACS Catal. **6**(10), 7133–7139 (2016)
75. Weekes, D.M., et al.: Electrolytic CO_2 reduction in a flow cell. Acc. Chem. Res. **51**(4), 910–918 (2018)
76. Seger, B., et al.: Best practices for electrochemical reduction of carbon dioxide. Nat. Sustain. **6**(3), 236–238 (2023)
77. Sun, Z., et al.: Fundamentals and challenges of electrochemical CO_2 reduction using two-dimensional materials. Chem. **3**(4), 560–587 (2017)
78. Burdyny, T., Smith, W.A.: CO_2 reduction on gas-diffusion electrodes and why catalytic performance must be assessed at commercially-relevant conditions. Energy Environ. Sci. **12**(5), 1442–1453 (2019)
79. dos Reis, E.A., et al.: Revisiting electrocatalytic CO_2 reduction in nonaqueous media: promoting CO_2 recycling in organic molecules by controlling H_2 evolution. Energ. Technol. **11**(6), 2201367 (2023)
80. Liu, K., et al.: Introductory guide to assembling and operating gas diffusion electrodes for electrochemical CO_2 reduction. ACS Energy Lett. **4**(3), 639–643 (2019)
81. Iglesias van Montfort, H.-P., et al.: An advanced guide to assembly and operation of CO_2 electrolyzers. ACS Energy Lett. **8**(10), 4156–4161 (2023)
82. Chen, X., et al.: Electrochemical CO_2-to-ethylene conversion on polyamine-incorporated Cu electrodes. Nat. Catal. **4**(1), 20–27 (2021)
83. Ma, W., et al.: Electrocatalytic reduction of CO_2 and CO to multi-carbon compounds over Cu-based catalysts. Chem. Soc. Rev. **50**(23), 12897–12914 (2021)
84. Lu, X., et al.: In situ observation of the pH gradient near the gas diffusion electrode of CO_2 reduction in alkaline electrolyte. J. Am. Chem. Soc. **142**(36), 15438–15444 (2020)
85. García de Arquer, F.P., et al.: CO_2 electrolysis to multicarbon products at activities greater than 1 A cm^{-2}. Science **367**(6478), 661–666 (2020)
86. Gong, Q., et al.: Structural defects on converted bismuth oxide nanotubes enable highly active electrocatalysis of carbon dioxide reduction. Nat. Commun. **10**(1), 2807 (2019)
87. Bui, J.C., et al.: Engineering catalyst–electrolyte microenvironments to optimize the activity and selectivity for the electrochemical reduction of CO_2 on Cu and Ag. Acc. Chem. Res. **55**(4), 484–494 (2022)
88. Dinh, C.-T., et al.: CO_2 electroreduction to ethylene via hydroxide-mediated copper catalysis at an abrupt interface. Science **360**(6390), 783–787 (2018)
89. Zhang, X., et al.: Selective and high current CO_2 electro-reduction to multicarbon products in near-neutral KCl electrolytes. J. Am. Chem. Soc. **143**(8), 3245–3255 (2021)
90. Chen, C., et al.: Address the "alkalinity problem" in CO_2 electrolysis with catalyst design and translation. Joule. **5**(4), 737–742 (2021)

91. Cao, Y., et al.: Surface hydroxide promotes CO_2 electrolysis to ethylene in acidic conditions. Nat. Commun. **14**(1), 2387 (2023)
92. Wang, Z., et al.: Localized alkaline environment via in situ electrostatic confinement for enhanced CO_2-to-ethylene conversion in neutral medium. J. Am. Chem. Soc. (2023)
93. Ma, W., et al.: Electrocatalytic reduction of CO_2 to ethylene and ethanol through hydrogen-assisted C-C coupling over fluorine-modified copper. Nat. Catal. **3**(6), 478–487 (2020)
94. Merino-Garcia, I., et al.: Tailoring gas-phase CO_2 electroreduction selectivity to hydrocarbons at Cu nanoparticles. Nanotechnology **29**(1), 014001 (2017)
95. Zhang, Y., et al.: Cu/Cu_2O nanoparticles supported on vertically ZIF-L-coated nitrogen-doped graphene nanosheets for electroreduction of CO_2 to ethanol. ACS Appl. Nano Mater. **3**(1), 257–263 (2019)
96. Ding, L., et al.: Over 70% Faradaic efficiency for CO_2 electroreduction to ethanol enabled by potassium dopant-tuned interaction between copper sites and intermediates. Angew. Chem. Int. Ed. **61**(36), e202209268 (2022)
97. Xu, H., et al.: Highly selective electrocatalytic CO_2 reduction to ethanol by metallic clusters dynamically formed from atomically dispersed copper. Nat. Energy **5**(8), 623–632 (2020)
98. Geioushy, R., et al.: Graphene/ZnO/Cu_2O electrocatalyst for selective conversion of CO_2 into n-propanol. Electrochim. Acta **245**, 456–462 (2017)
99. Liu, Y., et al.: Efficient electrochemical reduction of carbon dioxide to acetate on nitrogen-doped nanodiamond. J. Am. Chem. Soc. **137**(36), 11631–11636 (2015)
100. Guo, W., et al.: Highly efficient CO_2 electroreduction to methanol through atomically dispersed Sn coupled with defective CuO catalysts. Angew. Chem., Int. Ed. **60**(40), 21979–21987 (2021)
101. Qiu, Y.-L., et al.: Copper electrode fabricated via pulse electrodeposition: toward high methane selectivity and activity for CO_2 electroreduction. ACS Catal. **7**(9), 6302–6310 (2017)
102. Hou, X., et al.: 3D core–shell porous-structured Cu@ Sn hybrid electrodes with unprecedented selective CO_2-into-formate electroreduction achieving 100%. J. Mater. Chem. A. **7**(7), 3197–3205 (2019)
103. Zhong, M., et al.: Accelerated discovery of CO_2 electrocatalysts using active machine learning. Nature **581**(7807), 178–183 (2020)
104. Liu, W., et al.: Electrochemical CO_2 reduction to ethylene by ultrathin CuO nanoplate arrays. Nat. Commun. **13**(1), 1877 (2022)
105. Zhao, Y., et al.: Industrial-current-density CO_2-to-C_{2+} electroreduction by anti-swelling anion-exchange ionomer-modified oxide-derived Cu nanosheets. J. Am. Chem. Soc. **144**(23), 10446–10454 (2022)
106. Gu, Z., et al.: Efficient electrocatalytic CO_2 reduction to C_{2+} alcohols at defect-site-rich Cu surface. Joule. **5**(2), 429–440 (2021)
107. Fang, M., et al.: Metalloporphyrin-linked mercurated graphynes for ultrastable CO_2 electroreduction to CO with nearly 100% selectivity at a current density of 1.2 A cm^{-2}. J. Am. Chem. Soc. **144**(33), 15143–15154 (2022)
108. Cuellar, N.R., et al.: Advantages of CO over CO_2 as reactant for electrochemical reduction to ethylene, ethanol and n-propanol on gas diffusion electrodes at high current densities. Electrochim. Acta **307**, 164–175 (2019)
109. Fan, L., et al.: Electrochemical CO_2 reduction to high-concentration pure formic acid solutions in an all-solid-state reactor. Nat. Commun. **11**(1), 3633 (2020)
110. Bagchi, D., et al.: Structure-tailored surface oxide on Cu–Ga intermetallics enhances CO_2 reduction selectivity to methanol at ultralow potential. Adv. Mater. **34**(19), 2109426 (2022)
111. Zhang, L., et al.: Enhanced cuprophilic interactions in crystalline catalysts facilitate the highly selective electroreduction of CO_2 to CH_4. J. Am. Chem. Soc. **143**(10), 3808–3816 (2021)

112. Peng, C., et al.: Highly-exposed single-interlayered Cu edges enable high-rate CO_2-to-CH_4 electrosynthesis. Adv. Energy Mater. **12**(15), 2200195 (2022)
113. Li, J., et al.: Silica-copper catalyst interfaces enable carbon-carbon coupling towards ethylene electrosynthesis. Nat. Commun. **12**(1), 2808 (2021)
114. Shang, L., et al.: Efficient CO_2 electroreduction to ethanol by Cu_3Sn catalyst. Small Methods. **6**(2), 2101334 (2022)
115. Xu, Y., et al.: Low coordination number copper catalysts for electrochemical CO_2 methanation in a membrane electrode assembly. Nat. Commun. **12**(1), 2932 (2021)
116. Wu, Y., et al.: Enhancing CO_2 electroreduction to CH_4 over Cu nanoparticles supported on N-doped carbon. Chem. Sci. **13**(28), 8388–8394 (2022)
117. Jeong, H.-Y., et al.: Achieving highly efficient CO_2 to CO electroreduction exceeding 300 mA cm^{-2} with single-atom nickel electrocatalysts. J. Mater. Chem. A. **7**(17), 10651–10661 (2019)
118. Ge, L., et al.: Electrochemical CO_2 reduction in membrane-electrode assemblies. Chem. **8**(3), 663–692 (2022)
119. Gabardo, C.M., et al.: Continuous carbon dioxide electroreduction to concentrated multi-carbon products using a membrane electrode assembly. Joule. **3**(11), 2777–2791 (2019)
120. Ma, S., et al.: Carbon nanotube containing Ag catalyst layers for efficient and selective reduction of carbon dioxide. J. Mater. Chem. A. **4**(22), 8573–8578 (2016)
121. Lu, X., et al.: Electrochemical reduction of carbon dioxide to formic acid. ChemElectroChem **1**(5), 836–849 (2014)
122. Ripatti, D.S., et al.: Carbon monoxide gas diffusion electrolysis that produces concentrated C_2 products with high single-pass conversion. Joule. **3**(1), 240–256 (2019)
123. Lv, J.J., et al.: A highly porous copper electrocatalyst for carbon dioxide reduction. Adv. Mater. **30**(49), 1803111 (2018)
124. Wu, Y., et al.: Mitigating electrolyte flooding for electrochemical CO_2 reduction via infiltration of hydrophobic particles in a gas diffusion layer. ACS Energy Lett. **7**(9), 2884–2892 (2022)
125. Hansen, K.U., Jiao, F.: Hydrophobicity of CO_2 gas diffusion electrodes. Joule. **5**(4), 754–757 (2021)
126. Xing, Z., et al.: Beyond catalytic materials: controlling local gas/liquid environment in the catalyst layer for CO_2 electrolysis. J. Energy Chem. **66**, 45–51 (2022)
127. Xing, Z., et al.: Enhancing carbon dioxide gas-diffusion electrolysis by creating a hydrophobic catalyst microenvironment. Nat. Commun. **12**(1), 136 (2021)
128. Li, M., et al.: The role of electrode wettability in electrochemical reduction of carbon dioxide. J. Mater. Chem. A. **9**(35), 19369–19409 (2021)
129. Bi, J., et al.: Construction of 3D copper-chitosan-gas diffusion layer electrode for highly efficient CO_2 electrolysis to C_{2+} alcohols. Nat. Commun. **14**(1), 2823 (2023)

Enhanced CO_2-to-CH_4 Conversion via Grain Boundary Oxidation Effect in CuAg Systems

3

3.1 Introduction

ECR into value-added chemicals and fuels using renewable resources helps tackle the challenges of energy crisis and global warming, but this strategy highly depends on the design and fabrication of electrocatalysts with great activity and selectivity to minimize the energy consumption [1–4]. Among the state-of-the-art ECR catalysts, copper is the only metal capable of reducing CO_2 into hydrocarbons and alcohols [5]. Therefore, numerous approaches, including alloying [6, 7], oxide-derived copper (OD-Cu) [8, 9], grain boundaries (GBs) [10, 11], morphology modification [12, 13], and interface and strain engineering [14, 15], have been developed to increase the capability of the Cu-based catalysts. The main technique of these approaches is to create specific ensembles with controllable selectivity towards the targeted products (e.g. ethylene and methane). Nevertheless, the structure-performance relationship between electrocatalysts and the activity and selectivity of ECR remain unclear, hindering the development of Cu-based catalysts [4].

Among the preceding strategies for catalyst design, GBs are widely used to tune the activity and selectivity of ECR [10, 16]. The GBs, as one of structural defects, can create substantial structural perturbations in their vicinity, strained regions, and high-energy surfaces for catalysis [17]. Thus, one of the essential tasks in catalyst design is to enhance the density of GBs to manipulate each mechanistic step of the ECR reactions. The GBs generated from OD-Cu have attracted significant attention among the previous preparation protocols [8, 9]. During in situ reduction of oxide/hydroxide crystals into nanosized irregular Cu grains, this method can be used to produce rich GBs accounting for the enhanced selectivity of C_{2+} products, such as ethylene and ethanol [8]. Moreover, previous studies indicate that the density of the GBs on the surface of OD-Cu directly correlates with the

L. Wang et al., *Modulation Strategies of Cu-based Electrocatalysts for Enhancing Electrocatalytic CO_2 Conversion*, Synthesis Lectures on Green Energy and Technology,
https://doi.org/10.1007/978-3-031-98056-5_3

binding energy of the key intermediate (e.g., $^{*}CO$), greatly expediting the reactions and promoting C–C coupling [10, 18].

According to the Pourbaix diagram of Cu metal, oxidized Cu precursors should be completely reduced to metallic Cu(0) when negative potentials are applied [19]. However, the presence of $Cu^{\delta+}$ species and copper oxide has been reported under ECR conditions [20, 21]. These earlier experimental results suggest that these species are persistent during ECR, acting as active sites for high C_{2+} selectivity [22]. Nonetheless, other experimental studies show that only metallic Cu exists at the reduction potential in ECR [23, 24]. These conflicting statements in the literature indicate the difficulty to accurately identify the active sites of OD-Cu in ECR.

To date, although the grain boundaries was considered to tune the electrocatalytic activity and selectivity of OD-Cu [25], the GBO effect remains unclear in the OD-Cu system. Highly active GBs at the small Cu grains are usually unstable and liable to oxidation, although high-density GBs on the catalyst surface are desired [26, 27]. For example, the uncoordinated Cu atoms on the GBs in Cu–Ag thin-film are susceptible to spontaneous oxidation in the ambient air, but the oxidized GBs degrade the Faradaic efficiency (FE) of CO and CH_4 in neutral electrolyte [28]. However, quantum mechanics indicates that the borders of the oxidized and metallic Cu surface regions in the oxidized matrix significantly enhances the ECR activity [29]. Although OD-Cu catalyst system are generally believed to promote C–C coupling toward C_2 products [20–22], this mechanism does not apply to the reaction system of C_1 products like CO and CH_4, especially in the Cu-based bimetallic alloy catalysts. Incorporating oxygen into the GBs in bimetallic CuAg alloys further complicates the alloy system because of the compatibility of Cu_2O, CuO and Ag_2O [30]. Thus, the conflicting conclusion reported in the literature indicate the challenges to understanding the role of GB effect and oxidation states on the activity and selectivity of ECR to CH_4 in pH-neutral electrolyte.

This work presents an electro-shock (ES) reduction strategy for the synthesis of bimetallic nanoparticles. By ES reduction, three distinct forms of CuAg nanoparticles with unique structural features (such as GBs and GBO) exhibit different ECR activities, highlighting the superiority of GBO structure. The incorporation of oxidation effect into GBs in the ES-1%CuAg significantly improves the FE of CH_4, which is twice higher than that of ES-Cu or ES-50%CuAg. We have unraveled the mechanism underlying the GBO effect during ECR using in situ X-ray absorption spectroscopy (XAS) and ATR FT-IR spectra. Our findings highlight the role of stabilized $Cu^{\delta+}$ species as active sites in ECR, and the unique electron transfer mechanism facilitated by Cu–O bridge at GBs rather than the Cu–Ag bond. DFT calculations further validate the significance of weak $^{*}CO$ adsorption in promoting the conversion of ECR to CH_4. These deep understanding of GBO effect helps clarify the conflicting conclusions in the literature, and sheds light on the design and synthesis of GBO structure for effective ECR.

3.2 Experimental and Computational Methods

3.2.1 Materials

Copper(II) chloride dihydrate ($CuCl_2 \cdot 2H_2O$), silver nitrate ($AgNO_3$), 60–80 nm Cu nanoparticles and potassium bicarbonate ($KHCO_3$) were purchased from Sigma-Aldrich (Canada). Nitric acid (HNO_3), ethanol, ultrapure water (18 MΩ cm), and iso-propanol were obtained from the chemical store on the University of Waterloo campus. Carbon cloth without microporous layer (MPL) was purchased from fuel cell store, USA. Argon and CO_2 gas were purchased from Linde Gas, Canada.

3.2.2 Synthesis of CuAg Nanoparticles

An electro-shock method was developed to synthesize CuAg nanoparticles following these steps. First, $0.5 \times 1\ cm^2$ carbon cloth was activated in a 100 mL of 2 M nitric acid (HNO_3) solution with constant stirring at 120 °C for 2h. After activation, the resulting carbon cloth was successively washed three times with water and ethanol until reaching constant pH, and dried for later use. Second, for the preparation of the 10 mM metal precursor solution, $CuCl_2$ and $AgNO_3$ were dissolved into ethanol and mixed solvent (water/ethanol = 1:3), respectively. Then the precursor solutions with targeted loading concentrations (Cu:Ag = 99:1, 100 μL total volume) were sequentially sprayed onto the activated carbon cloth. The samples were left to dry at room temperature before electric shock using a Metrohm autolab (PGSTAT204) with the 10 A booster as an external power source, with a current of up to 10 A. A 10 A current for 0.1 s under argon protective gas was selected as the reaction condition for synthesizing the ES-1%CuAg electrocatalysts. The ES-Cu, ES-Ag, and ES-50%CuAg were prepared using same method, with only the volume ration of Cu and Ag metal precursor solution being varied, such as ES-Cu for 100%Cu, ES-Ag for 100%Ag, ES-50%CuAg for Cu:Ag = 1:1.

For control experiment, WC-1%CuAg nanoparticles were prepared by a standard wet chemical method. First, 10 mg of commercial Cu nanoparticles were dispersed in 5 mL of ethylene glycol. Then, 0.5 mg of $AgNO_3$ was added to the above solution, which was then sonicated for 1 h. The resultant WC-1%CuAg nanoparticles were washed three times by water and ethanol, respectively, followed by drying for 12 h under vacuum at 60 °C.

3.2.3 Ex-Situ Material Characterization

Scanning electron microscopy (SEM, Hitachi S4800, Japan) images were conducted under a working accelerating voltage of 10 kV. Glancing-incidence X-ray diffraction (GIXRD) pattern were performed on a MRD diffractometer (PANalytical X'Pert Pro, Netherlands) with Cu Kα radiation (1.54 Å) at an incidence angle of 0.3°. X-ray photoelectron spectroscopy (XPS) measurements were carried out (Thermo-VG Scientific ESCALab 250, USA) using microprobes with a monochromatic Al Kα X-ray source (1486.6 eV). High-angle annular dark-field scanning TEM (HAADF-STEM) and EDS images (FEI Titan 80–300 HB TEM/STEM, USA) were taken with an aberration corrector operating at 60 kV accelerating voltage. High resolution TEM images (Zeiss Libra 200MC, German) were collected under an acceleration voltage of 200 kV. Inductively coupled plasma mass spectrometry (ICP-MS) analysis was performed on an Agilent 8800 triple quadrupole, using He as a collision cell gas, Ge and In as internal standards to calibrate for instrument drift and ICP elemental standards to verify the accuracy of the instrument. Proton nuclear magnetic resonance (H-NMR) was run on Bruker Avance III 500 MHz, USA. The reduced products were evaluated on a gas chromatograph (Agilent 6890, USA) with a Carbonxen® 1000 column and a CarbonPLOT column for flame ionization detector (FID) and thermal conductivity detector (TCD), respectively.

3.2.4 Electrochemical Measurements

All electrochemical tests were conducted on an electrochemical workstation (Gamry Reference 3000, USA) with IR compensation at room temperature. The customized H-type cell composed of two compartments separated by the Nafion 211 membrane. An Ag/AgCl reference electrode, Pt counter electrode, and 0.5×1 cm^2 carbon cloth served as reference electrode, counter electrode and working electrode, respectively. The 0.1 M $KHCO_3$ electrolyte was saturated by bubbling CO_2 for 30 min before test. The ECR catalytic activities were evaluated under an applied cathodic potential for 1000 s with 10 sccm of flowing pure CO_2 gas. The gas products were characterized by the Agilent 6890. The liquid products were characterized with a Bruker 500 MHz Nuclear Magnetic Resonance (NMR) using an internal reference prepared with D_2O and DMSO.

All potentials in this work were measured against the Ag/AgCl reference electrode and converted into reversible hydrogen potential (RHE) using the equation below:

$$E\ (\text{vs. RHE}) = E\ (\text{vs. Ag/AgCl}) + 0.059 \times \text{pH} + 0.197$$

All the potentials are relative to the RHE unless stated otherwise.

3.2.5 In Situ XAS Measurements

The in situ XAS measurements were conducted at 06ID-1 Hard x-ray MicroAnalysis (HXMA) beamline of the Canadian Light Source (CLS) operated at 2.9 GeV with a constant current of 250 mA. The measurements at the copper K-edge and Ag K-edge were performed in fluorescence mode using a Lytle detector. The ECR was carried out in

CO_2-saturated 0.1 M $KHCO_3$ solution with 10 sccm of flowing pure CO_2 gas in a home-made cell. In situ XAS measurements was recorded at applied cathodic potential from 0 to − 1.1 V versus RHE. The XAS data were fitting using the software package Athena. During the fitting process, the Δk and ΔR ranges were set to 0–12.3 and 1.07–3.35, respectively, and S_0 was fixed at 0.78.

3.2.6 In Situ Time-Dependent EPR Measurement

Time-dependent continuous-wave (CW) EPR measurements at X-band frequencies (9.4 GHz) were performed using Bruker EMX Micro EPR spectrometer. DMPO was used as a spin trap. Electrochemistry procedures were performed with ES-1%CuAg in the CO_2-saturated 0.1 M $KHCO_3$ electrolyte containing 10 mM of DMPO. During ECR reaction, 10 μL of samples were taken from the electrochemical cell and transferred into 1 mm ID, 2 mm OD EPR tube (712-SQ-100M Wilmad LabGlass). For the EPR measurements these tubes were placed into 3 mm ID 4 mm OD EPR tubes (707-SQ-250M Wilmad LabGlass). For all the measurements microwave power was set to 1.262 mW (22 dB microwave power attenuation) and modulation amplitude was 2 G at 100 kHz modulation frequency. Simulations of the EPR spectra were performed with EasySpin software [31].

3.2.7 In Situ ATR FT-IR Measurements

The measurements were conducted at 02B1-1 Far-IR beamline of the Canadian Light Source (CLS). In situ ATR FT-IR were run using a Bruker IFS 125 HR spectrometer with a 9.4 m OPD with a custom-built Veemax III variable angle reflectance system (Pike instruments), and all spectra were given by the absorbance [− log(R/R_0)]. The catalyst ink was dropped via pipette onto an ITO-coated SI Microgrooved Wafer Element with a 55° Face angle and then left to dry in air before test. A platinum wire and an Ag/AgCl electrode were used as counter and reference electrodes, respectively. The 0.1M $KHCO_3$ electrolyte was constantly purged with CO_2 during the measurement. The applied cathodic potentials were changed from − 0.1 to − 1.7 V, and the infrared spectra were collected with a time resolution of 2 min per spectrum.

3.2.8 Computational Details

All spin-polarized DFT calculations were performed within the Vienna Ab initio Simulation Package (VASP) using the projector augmented wave method [32]. The generalized gradient approximation (GGA) of Perdew-Burke-Ernzerhof (PBE) was used to describe the exchange and correlation effects [33]. A kinetic energy cutoff of 500 eV was used to set the plane-wave basis set. A Gaussian smearing of the width of 0.05 eV was used during calculations. Geometrical optimizations stopped when the energy convergence was less than 10^{-6} eV and the forces on each atom were less than 0.005 eV/Å. The $3 \times 1 \times 1$ and $9 \times 3 \times 1$ Monkhorst– Pack k-point meshes were used for geometrical optimizations and electronic properties calculations, respectively. The van der Waals interactions were corrected by the Grimme's DFT-D3 method [34]. A vacuum space of 15 Å was added along the z axis to avoid the interlayer interactions. The free energy (ΔG) was determined by $\Delta G = \Delta E + \Delta ZPE - T\Delta S$ [35], where ΔE is the reaction energy, ΔZPE is the change of zero-point energy, T is the temperature (298.15 K), and ΔS is the change of entropy.

3.3 Results

3.3.1 Synthesis and Ex-Situ Characterization of Electrocatalyst

The metal-precursor solution is sprayed sequentially onto an activated carbon cloth, which is subject to the Joule heating created by 0.1 s of electric shock. The thermal shock breaks down the metal salts [$CuCl_2$ and $AgNO_3$] into Cu and Ag atoms in only 0.1s. The rapid thermal shock prevents Cu and Ag atoms from diffusing to form an equilibrium phase. However, the surface defects and oxygen-containing functional groups in the activated carbon cloth create kinetic anchoring sites. As a result, high-temperature mixing entropy drives the Cu and Ag atoms to mix, forming non-equilibrium bimetallic nanoparticles (Fig. 3.1a).

The scanning electron microscope (SEM) images (Figs. 3.1b, c and A.1-3) reveal the resultant in situ growth of many spherical CuAg nanoparticles on the activated carbon cloth. In a typical synthesis process, lowering the concentration of the Cu metal precursor solution greatly reduces the surface distribution and sizes of the spherical Cu nanoparticles on the carbon cloth because of insufficient Cu metal sources (Fig. A.1a–d). However, the size of ES-Ag nanoparticles (Fig. A.2a, b) are significantly larger than the ES-Cu nanoparticles (Fig. A.2c, d). The same reaction conditions with different metallic elements indicate that Ag atoms tend to aggregate more. As shown in Fig. A.3a, b, the nanoparticles formed at 0.05 s of electric shock are relatively small with extremely low surface coverage, indicating that 0.05 s of electric shock is insufficient for adequate migration and nucleation of metal atoms. However, extending the electro-shock time (> 0.1s) promotes the nucleation and overgrowth of nanoparticles, resulting in a decrease in surface distribution (Fig. A.3c–f). Based on the experimental results, 0.1 s of electric shock is the optimal for synthesizing bimetallic CuAg nanoparticles. After the introduction of Ag into Cu matrix (Figs. 3.1b, c and A.4a–f), the ES-1%CuAg (Fig. A.4c, d) and ES-50%CuAg (Fig. A.4e, f) exhibit much higher surface dispersibility than that of ES-Cu (Fig. A.4a, b), indicating that adding silver significantly enhances the nucleation rate of

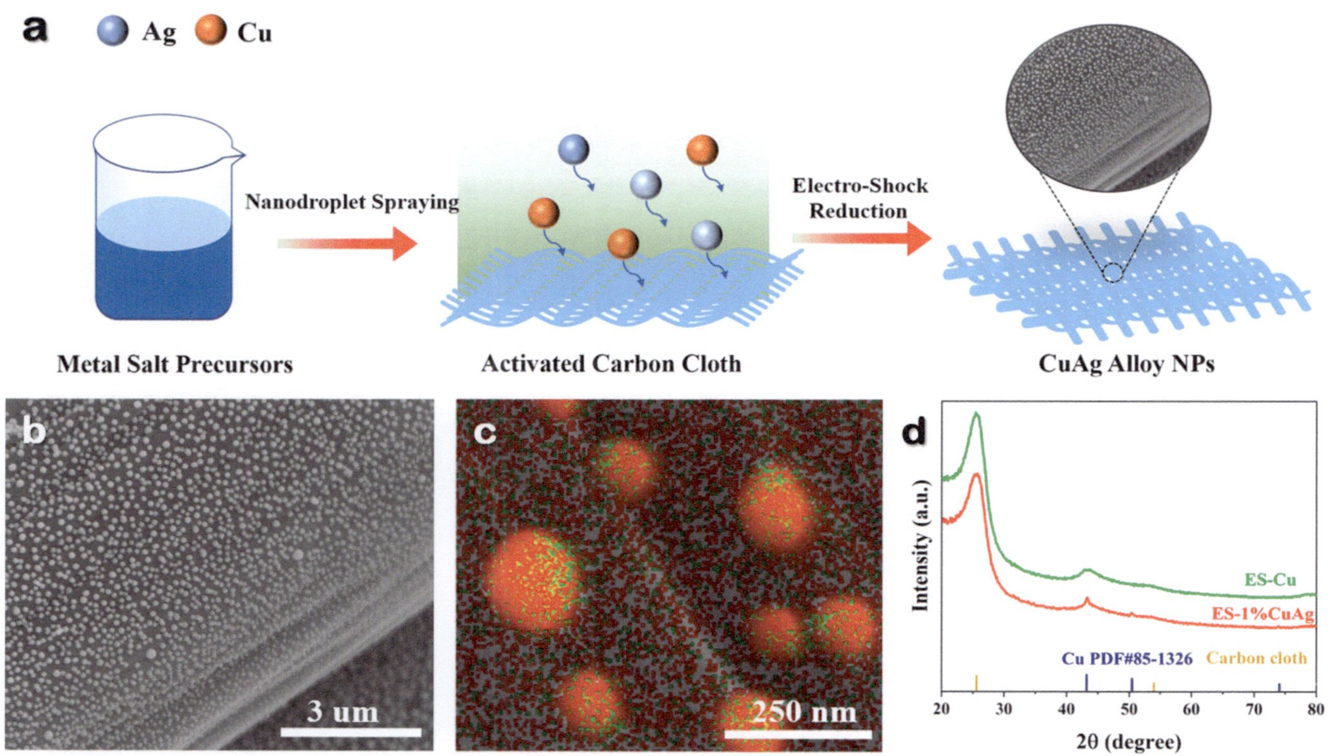

Fig. 3.1 **a** Scheme of the synthesis processes of CuAg catalysts by electro-shock method. **b** SEM image and **c** corresponding EDS element mapping image of ES-1%CuAg sample. **d** GIXRD spectra of ES-Cu and ES-1%CuAg

nanoparticles, even at low concentrations, while excessive Ag loading would cause the CuAg nanoparticles to grow larger (Fig. A.4c, d vs. A.4e, f).

The composition and structure features of the prepared samples were further studies using grazing incidence X-ray diffraction (GIXRD). No Ag reflection is observed in GIXRD pattern because of the low Ag loading and sample content on the carbon cloth, and the strong background signal (25.5° and 54°) from the carbon cloth support results in a large baseline that affects the quality of the Cu signals (Fig. 3.1d). Compared to the GIXRD patterns of Cu nanoparticles and WC-1%CuAg nanoparticles prepared by wet chemistry (WC) method (Methods, Fig. A.5), the absence of Cu_2O peaks at 36.3° and 61.2° (Fig. 3.1d) indicates that solid-solution structure and entropy effect causes the ES-Cu and ES-1%CuAg to exhibit a different atomic arrangement from that of the WC-1%CuAg. The ICP-MS results show that the Ag content in ES-1%CuAg and ES-50%CuAg is 1.7 wt% and 47.5 wt% (Table A.1), respectively. The aberration corrected-high-angle annular dark-field scanning transmission electron microscopy (AC-HAADF-STEM) (Fig. 3.2a, b) and high-resolution transmission electron microscopy (HRTEM) (Fig. A.6a–c) images reveal the existence of GBs in all samples. Notably, the ES-Cu also exhibits the GBs structure (Fig. A.6a), indicating that solid-solution structure induced by the electro-shock method is a non-equilibrium phase, and the metal atoms are randomly distributed. ES-1%CuAg (Fig. 3.2a) and ES-50%CuAg (Fig. 3.2b) show a face-centered cubic (fcc) crystalline structure with $d_{(111)}$ spacing of 0.22 and 0.23 nm between the spacing values of Cu and Ag metals, respectively [36]. Energy-dispersive x-ray spectroscopy (EDS) elemental mapping images reveal the rich accumulation of Ag and oxygen atoms on the GBs, as well as bulk distribution of oxygen atoms across grains, in the ES-1%CuAg (Figs. 3.2c and A.7a). This indicates that the Ag and oxygen enrichment on the GBs allows for the incorporation of oxygen into the defect structure, subsequently triggering the oxidation of unsaturated Cu sites. In contrast, the homogeneous distribution of Cu and Ag atoms and the absence of oxygen atoms in the ES-50%CuAg suggest that the nanoalloy effect prevents Cu sites on GBs from oxidation (Figs. 3.2d and A.7b). As a result, the non-uniform distribution of Ag atoms in Cu matrix for the ES-1%CuAg serves as a prerequisite for the oxidation of Cu sites on the GBs.

The surface compositions and chemical states of the as-prepared catalysts were further characterized using X-ray photoelectron spectroscopy (XPS) (Figs. 3.3a and A.8a–f). The Cu $2p_{3/2}$ spectra of all samples can be deconvoluted into two peaks, corresponding to Cu^+/Cu^0 and Cu^{2+} peaks, respectively. The intensity ratio of Cu^{2+} to Cu^+/Cu^0 peaks for ES-1%CuAg is 11.45, which is higher than those of ES-Cu (2.08) and ES-50%CuAg (0.74) (Fig. 3.3a). This indicates that under the GBO effect, the surface copper sites undergo further oxidation when exposed to air during ex-situ characterization. Moreover, the peak position of Cu^+/Cu^0 shifts towards a higher binding energy from 932.4 eV (ES-Cu) to 932.6 eV (ES-1%CuAg) and 932.8 eV (ES-50%CuAg) (Fig. 3.3a), suggesting that the interaction between the Cu and Ag sites affects the surface electronic structure of the Cu atoms [37]. In contrast, for WC-1%CuAg, the absence of Cu^+/Cu^0 signal and

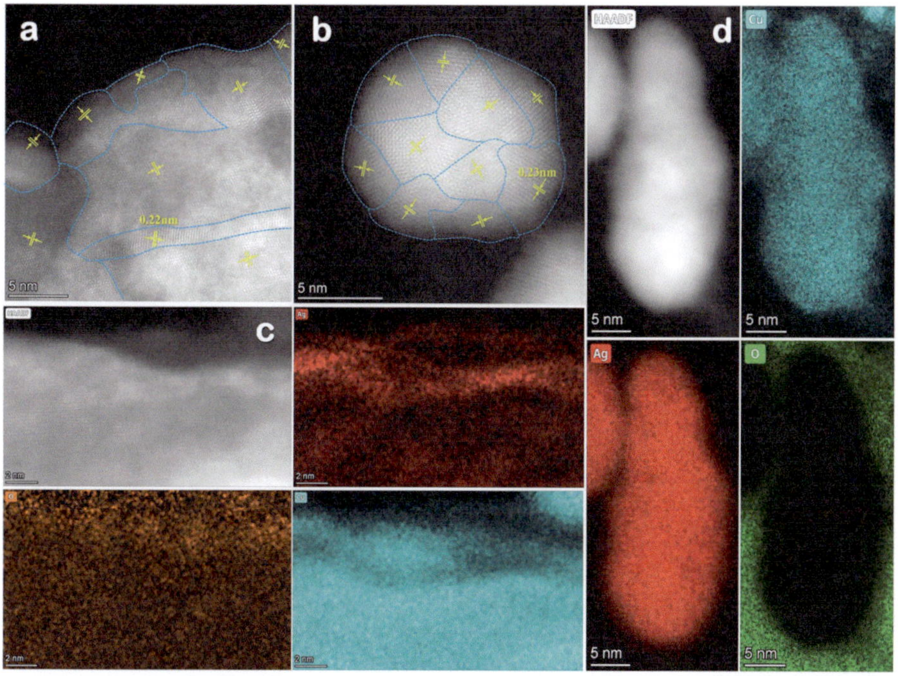

Fig. 3.2 AC-HAADF-STEM images of **a** ES-1%CuAg and **b** ES-50%CuAg. STEM-EDX element mapping image of **c** ES-1%CuAg and **d** ES-50%CuAg

the unshifted binding energy of Cu^{2+} indicate a stronger electron attraction effect of Ag in equilibrium phase (Fig. A.8a). Regarding O 1s XPS spectra, lattice oxygen observed in Cu nanoparticles and WC-1%CuAg is not present in ES-Cu, ES-1%CuAg and ES-50%CuAg (Fig. A.8b, c). Notably, the Cu–O bond of ES-1%CuAg shifts to a position with higher binding energy (532.4 eV) compared to ES-Cu (531.7 eV) and ES-50%CuAg (531.8 eV) (Fig. A.8c). This indicates that the GBs in ES-1%CuAg readily bind with oxygen, inducing stronger oxidation effect. Due to the inert atmosphere during synthesis, the source of oxygen can be attributed to the oxygen-containing groups on the activated carbon cloth (Fig. A.8c, d). Therefore, the oxygen-containing groups not only provide anchoring sites, but also are involved in the nucleation and growth of nanoparticles by bonding to unsaturated Cu sites. Compared to the binding energy of Ag 3d at 368.4 eV for ES-1%CuAg and ES-50%CuAg (Fig. A.8e), the binding energy of Ag 3d for WC-1%CuAg shifts to 369.5 eV (Fig. A.8f), indicating that the non-equilibrium phase exhibits completely different electronic structure and charge distribution between Cu and Ag from equilibrium phase [37].

Because the XPS is insufficient to accurately determine the oxidation state and local coordination environment of the Cu and Ag species under high dispersion and low

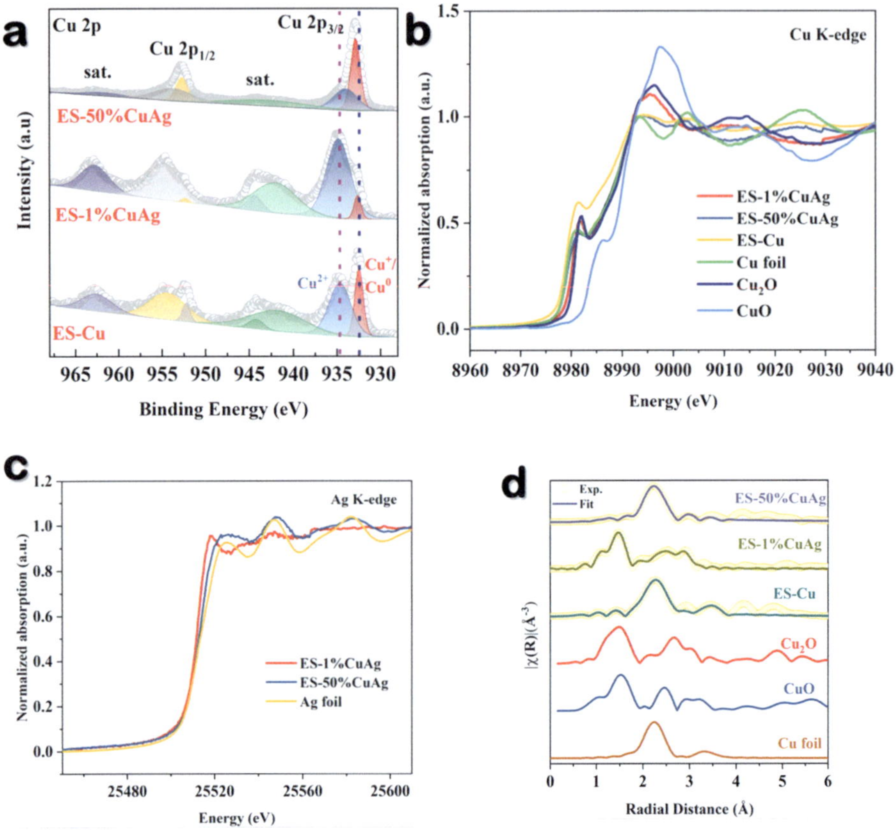

Fig. 3.3 **a** Cu 2p XPS spectra of ES-Cu, ES-1%CuAg and ES-50%CuAg catalysts; **b** Cu K-edge XANES spectra of ES-Cu, ES-1%CuAg and ES-50%CuAg samples and corresponding references (Cu foil, Cu_2O, and CuO); **c** Ag K-edge XANES spectra of ES-1%CuAg and ES-50%CuAg and corresponding references (Ag foil); **d** Cu K-edge EXAFS experimental and fitting spectrum and of ES-Cu, ES-1%CuAg and ES-50%CuAg catalysts and corresponding references (Cu foil, Cu_2O, CuO)

loading, synchrotron-based X-ray absorption Near-Edge Structure (XANES) spectra and Extended X-ray Absorption Fine Structure (EXAFS) spectra were performed to further investigate the GBO effect. Compared to the Cu foil and Cu_2O reference, the XANES spectra of ES-Cu and ES-50%CuAg exhibit characteristics of metallic Cu, while the Cu oxidation state in ES-1%CuAg is close to the Cu(I) state ($Cu^{\delta+}$, $0 < \delta < 1$) (Fig. 3.3b). The Ag loading significantly changes the valence state of Cu sites. Conversely, the Ag K-edge XANES spectrum of ES-1%CuAg is also completely different from that of Ag foil and ES-50%CuAg (Fig. 3.3c), indicating strong interaction between Ag and Cu. The coordination environment of Cu sites in all samples was further investigated through the structural parameters obtained from quantitative fitting of the EXAFS spectra (Figs. 3.3d, A.9 and Table A.2). The absence of Cu–O bond in ES-Cu and ES-50%CuAg further differentiate the presence of Cu–O bond induced via GBs oxidation effect in ES-1%CuAg. Compared to Cu foil, the similar Cu–Cu bond distance ($R_{Cu\text{–}Cu} = \sim 2.53$ Å) but significantly different Cu–Cu coordination number ($N_{Cu\text{–}Cu} = \sim 5.7$) in ES-Cu indicates that Cu sites in ES-Cu are in a lower coordination state (Table A.2). Since the catalyst prepared via electro-shock synthesis is a non-equilibrium phase with randomly mixed atoms, the catalyst is rich in GBs structures, leading to low-coordinated Cu sites. After silver loading, the $N_{Cu\text{–}Cu}$ sharply decreases from ~ 5.4 in ES-1%CuAg to ~ 1.9 in ES-50%CuAg, while the Cu–Ag coordination number ($N_{Cu\text{–}Ag}$) increases to ~ 4.2. This indicates that there is no a significant change in disorder of Cu–Cu after GBO, with the enhanced disorder after nanoalloying. Additionally, the introduction of silver extends the $R_{Cu\text{–}Cu}$ to ~ 2.87 Å in ES-1%CuAg and ~ 2.64 Å in ES-50%CuAg compared to ES-Cu, implying lattice expansion induced by Ag doping, while GBO effect leads to a greater lattice expansion.

3.3.2 In Situ Multimodal Characterization of Electrocatalysts for ECR in Neutral Media

In situ XAS was performed to reveal the structure evolution and charge transfer between Cu and Ag atoms during ECR. The in situ Cu K-edge XANES spectra of ES-1%CuAg in CO_2-saturated 0.1M $KHCO_3$ solution show a similar shape and position under open circuit potential (OCP) and cathodic potentials of -0.6 V_{RHE}, -0.9 V_{RHE} and -1.1 V_{RHE} (Fig. 3.4a and Figure A.10). Although the oxidation state of $Cu^{\delta+}$ species presents a downward trend (see inset in Fig. 3.4a), the $Cu^{\delta+}$ species are incompletely reduced to metallic Cu, even at high cathodic potentials. This indicates that the GBs structure introduced by electro-shock synthesis contributes to the structural stability and resistance to the electrochemical reduction under reduction potentials in ECR. Therefore, ES-Cu and ES-50%CuAg show the similar stability for $Cu^{\delta+}$ species during ECR (Figs. A.11a and A.12a). Additionally, despite the dominant Cu–Cu bond in ES-Cu and ES-50%CuAg before reaction, the metallic Cu undergoes rapid oxidation to $Cu^{\delta+}$ species under ECR reaction (Figs. A.11a and A.12a). This indicates that the in situ generated strongly oxidative species (e.g. OH radicals) lead to the oxidation of copper sites in the CO_2-saturated 0.1 M $KHCO_3$ electrolyte [38]. Consistent with the in situ Cu K-edge XANES spectra (Fig. 3.4a), the in situ Cu K-edge EXAFS spectra of ES-1%CuAg also show a significant Cu–O scattering path (Fig. 3.4b). With increasing cathodic potentials, the length of Cu–O bond increases (Fig. 3.4b). Similar trend can be observed in the Cu–O bond of ES-50%CuAg (Fig. A.12b, c), except for ES-Cu (Fig. A.11b, c).

To further expand the research, in situ Ag K-edge XANES spectra were collected (Figs. 3.4c and A.12d). The results obtained from ES-50%CuAg show electron transfer from electron-rich Ag to electron-deficient Cu during the ECR (Fig. A.12d). However,

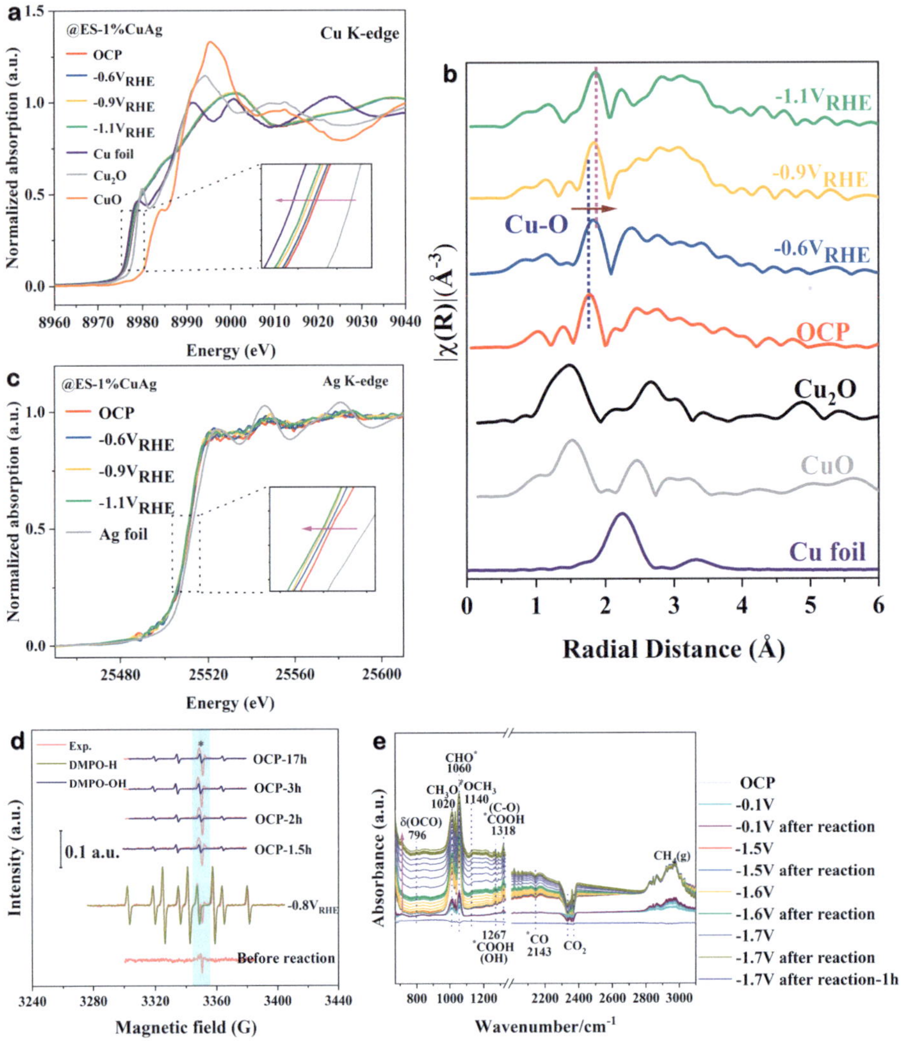

Fig. 3.4 **a** In situ Cu K-edge XANES spectra, **b** in situ Cu K-edge EXAFS spectra, and **c** in situ Ag K-edge XANES spectra of ES-1%CuAg under OCP and cathodic potentials from − 0.6 V_{RHE}, − 0.9 V_{RHE} to − 1.1 V_{RHE}; **d** in situ time-dependent normalized EPR spectra of 0.1 M $KHCO_3$ electrolyte at − 0.8 V_{RHE}, followed by switching to OCP for different periods on ES-1%CuAg sample. Experimental spectra are depicted in red, while simulated spectra are depicted in blue and dark yellow for DMPO-OH and DMPO-H radicals, respectively. EPR signals marked by asterisk is from the sample tube **e** In situ ATR FT-IR spectra collected from ES-1%CuAg catalyst at different applied potentials (− 0.1 V, − 1.5 V, 1.6 V and − 1.7 V)

regarding ES-1%CuAg, it is noteworthy that the Ag K-edge absorption spectra (Fig. 3.4c) still indicate a reducing effect similar to that observed in Cu K-edge spectra (Fig. 3.4a), which indicates that the electron transfer is not from Ag to Cu in the ES-1%CuAg. Such a different behavior between ES-50%CuAg and ES-1%CuAg can be attributed to the GBO effect in ES-1%CuAg. In the ES-1%CuAg, oxygen interacts with the surface electrons of the Cu and Ag atoms, driving electron transfer from electron-rich O to electron-deficient Cu and Ag. This electron transfer may alter the electronic structure and chemical properties of the Cu sites, enhancing ECR activity. Moreover, this mechanism behind electron transfer from O to Cu/Ag atoms in ES-1%CuAg significantly differs from that observed in ES-50%CuAg and those previously reported in the literature [37, 39].

To confirm the impact of strongly oxidative species in the CO_2-saturated 0.1 M $KHCO_3$ electrolyte on the rapid oxidation of active Cu sites during ECR, in situ time-dependent electron paramagnetic resonance (EPR) was conducted (Fig. 3.4d). 5,5-dimethyl-1-pyrroline N-oxide (DMPO) is used to prolong the lifetime of the radicals to the minute range. When the electrolyte contains 10 mM DMPO, an electrode potential of $-$ 0.8 V_{RHE} applied on ES-1%CuAg leads to hydrogen evolution reaction (HER) and subsequent trapping of hydrogen radicals ($H^{\cdot}$) in the form of a DMPO-H adduct (hyperfine splitting constants, A_N = 16.4 mT, A_H = 2.26 mT) [38]. Upon switching to OCP, OH radicals appear as a DMPO-OH adduct (A_N = 1.50 mT, A_H = 1.48 mT) and become dominant after 1.5 h, persisting throughout the OCP process [38]. This finding provides further evidence that, during the ECR reaction, the OH radicals generated from electrolyte excitation act as potent in situ oxidizing species, leading to the rapid oxidation of active Cu sites.

To detect the reaction intermediates during the ECR process, in situ attenuated total reflection Fourier transform infrared spectroscopy (ATR FT-IR) measurement was performed for ES-1%CuAg. The adsorption bands at 796 cm^{-1}, 1267 cm^{-1}, and 1318 cm^{-1} are assigned to δ(OCO), OH deformation, and C–O stretch, respectively (Fig. 3.4e). They are generally considered as the important intermediates during ECR to CO or CH_4 [40, 41]. The characteristic peak at 2143 cm^{-1} belongs to the adsorbed $^*CO$, which would undergo further proton-coupled electron transfer (PCET) process [42]. This evolution is further confirmed by the $^*CH_3O$ group adsorption bands at 1020 cm^{-1} and 1140 cm^{-1} and the peak of $^*CHO$ group at 1060 cm^{-1} (Fig. 3.4e); both of them are critical intermediates for ECR to CH_4 [43, 44]. Therefore, based on the in situ ATR FT-IR spectra analysis (Fig. 3.4e), the most likely reaction pathway for ECR to CH_4 are proposed as followed. Firstly, dissolved CO_2 in the liquid electrolyte is absorbed, activated on the catalytic sites, and transformed into $^*COOH$ by PCET process. Then, the hydrogen bonds-stabilized $^*COOH$ intermediate releases a water molecule to form $^*CO$ intermediate via PCET. Subsequently, $^*CO$ intermediate is still absorbed on the active sites by a hydrogen bond and undergoes a series of PCET processes to produce $^*CHO$ and $^*CH_3O$ intermediates, finally leading to selective production of CH_4.

In addition, the real-time ATR FT-IR spectra were used to monitor the intermediate adsorption processes at electrode–electrolyte interface. The in situ real-time ATR FT-IR spectra were collected every two minutes during the application of potentials, as well as after the potential was turned off, in a CO_2-staturated 0.1 M $KHCO_3$ (Figs. 3.4e and A.13a–d). The intermediate adsorption bands during ECR to CH_4 could be clearly observed even when the applied cathodic potential is as low as − 0.1 V, indicating that − 0.1 V can trigger the ECR reaction (Fig. 3.4e). With the increasing potentials or prolonging reaction time, the solubility of CO_2 further increases in the electrolyte, enhancing the intensity of intermediate characteristic peaks (Fig. A.13a–d). Furthermore, it is deemed reasonable that intermediates adsorption during ECR is reversible, and the signals for intermediates promptly disappear with potential withdrawal. However, once potentials were off, all peaks in our system did not disappear intermediately (Figs. 3.4e and A.13a–d). Instead, in situ generated intermediates remained on the catalyst surface or in the electrolyte for one hour after reaction at − 1.7 V (Figs. 3.4e and A.13d). Combined with the in situ Cu K-edge and Ag K-edge XAS, these results indicate that in situ generated ECR intermediates adsorbed on the surface of ES-1%CuAg, and that active $Cu^{\delta+}$ species promote the chemical evolution of adsorbate during ECR. However, potential withdrawal stops the continuous evolution. The sluggish reversible process indicates that active sites in turn prolongs the temporary existence of intermediates in the electrolyte by hydrogen bond. Eventually, the intermediates either disappear or evolve into the final products.

3.3.3 Performance of ECR in Neutral Media

The ECR performance of the as-synthesized catalysts were evaluated. Comparative analysis reveals that ES-1%CuAg exhibits the highest current density among all the catalysts (Fig. A.14a), and the current density of ES-1%CuAg in CO_2-saturated 0.1 M $KHCO_3$ electrolyte is higher than that in Ar-saturated 0.1 M $KHCO_3$ electrolyte, indicating the catalytic activity of ECR (Fig. A.14b). In more detail, the major products are H_2, CO, CH_4, and C_2H_4 for all samples with minor formate and ethanol (Fig. 3.5a–d and A.15a, b). Similar volcano-shaped curves trends for CO, C_2H_4 and CH_4 FEs on a wide range of applied potentials can be observed for all the samples (Fig. 3.5a–c). In terms of CO production (Fig. 3.5a), adding Ag into Cu matrix could significantly improve CO FE by four times compared to ES-Cu, indicating that Ag does facilitate the CO production. Notably, there was no significant difference in ethylene selectivity across the entire potential window for all catalysts (Fig. 3.5b), indicating that our CuAg system does not favor C–C coupling towards ethylene formation. Therefore, ES-1%CuAg shows the highest activity against CH_4 (~ 50% FE at − 1.1V_{RHE}) (Fig. 3.5c, d), which is nearly twice the CH_4 FE observed in ES-Cu and ES-50%CuAg. This indicates that GBO effect (ES-1%CuAg) is stronger than GBs effect (ES-Cu) and nanoalloy effect (ES-50%CuAg) on promoting the CH_4 reaction pathway in our system. Additionally, a comparative analysis was further conducted to evaluate the effect of non-equilibrium (ES-Cu, ES-1%CuAg and ES-50%CuAg) and equilibrium phases (commercial Cu and WC-1%CuAg) on CH_4 and C_{2+} product selectivity (Figs. 3.5a–d and A.16a, b). Compared to the non-equilibrium phase, the equilibrium phase system significantly promoted the selectivity of ethanol and formate (Fig. A.16a, b). However, even with the introduction of Ag into the equilibrium phase system, it neither suppressed CH_4 formation nor enhanced the selectivity for ethylene and ethanol compared to commercial Cu NPs (Fig. A.16a vs. b). As reported in other literature, Ag–Cu surfaces rich in low-coordination sites and CuAg catalysts with unique electronic structures modulate the CO adsorption strength, selectively promoting CH_4 production [37, 45]. Therefore, we can conclude that in the CuAg system with rich GBs structure, the unique GBO structure is more favorable for CH_4 production, and the nanoalloy effect is not as effective in promoting the production of C_{2+} products in neutral electrolytes [1, 39]. This indicates the GBO structure facilitates the hydrogenation of *CO rather than C–C coupling via the tandem mechanism.

More specifically, the main reduced products of the as-synthesized non-equilibrium phase samples at $-$ 1.1V_{RHE} were compared (Fig. 3.5e). Similar C_2H_4 and CH_4 FEs for ES-Cu and ES-50%CuAg indicate that the nanoalloy effect does not effectively promote the selective production of C_2H_4. The CH_4/C_2H_4 ratio (i.e., 5.3) of ES-1%CuAg almost doubles that of ES-Cu (i.e., 2.62) and ES-50%CuAg (i.e., 2.7). Such a volcano-shaped dependence of CH_4/C_2H_4 ratio on the three samples highlights the superiority of GBO structure in CH_4 conversion. Moreover, the CH_4/C_2H_4 ratio of the equilibrium phase (~ 1.0 for commercial Cu and ~ 1.33 for WC-1%CuAg) is much lower than that of the non-equilibrium phase, indicating that the equilibrium phase exhibits similar tendencies for CH_4 and C_2H_4, while the unique GBO structure in the non-equilibrium phase helps break the linear scale relationship for C_1 and C_2 products, selectively promoting CH_4 conversion. Additionally, at $-$ 0.8 V_{RHE} and $-$ 0.9 V_{RHE}, the CO partial current density increases with increasing potentials, while after $-$ 1.0 V_{RHE}, the H_2 partial current density rises sharply, far exceeding the slowly increasing CO partial current density (Fig. A.17a–c). In contrast to ES-Cu and ES-50%CuAg, the CH_4 partial current density of ES-1%CuAg is significantly higher than that of H_2, indicating that the GBO effect can effectively suppress the HER. The slow increase in the CO partial current density, combined with the high partial current density of H_2 and CH_4, suggests that our CuAg system does not follow the one-pot tandem fashion proposed for Ag–Cu surfaces [45], where *CO produced by Ag migrates to Cu sites. Instead, CO_2 preferentially adsorbs at the low-coordinated Cu sites at the GBO structure, and the hydrogen-covered Cu surface drives the hydrogenation of *CO to form *CHO and *CH_3O intermediates, finally leading to CH_4 production.

The long-term stability test of ES-1%CuAg in an H-cell was evaluated at a current density of 40 mA cm^{-2}. The catalyst was able to operate stably for 17 h, with the selectivity for methane gradually decreasing from the initial 48% to 38%, demonstrating a certain degree of stability (Fig. 3.5f). After the stability test, nanoparticles remained

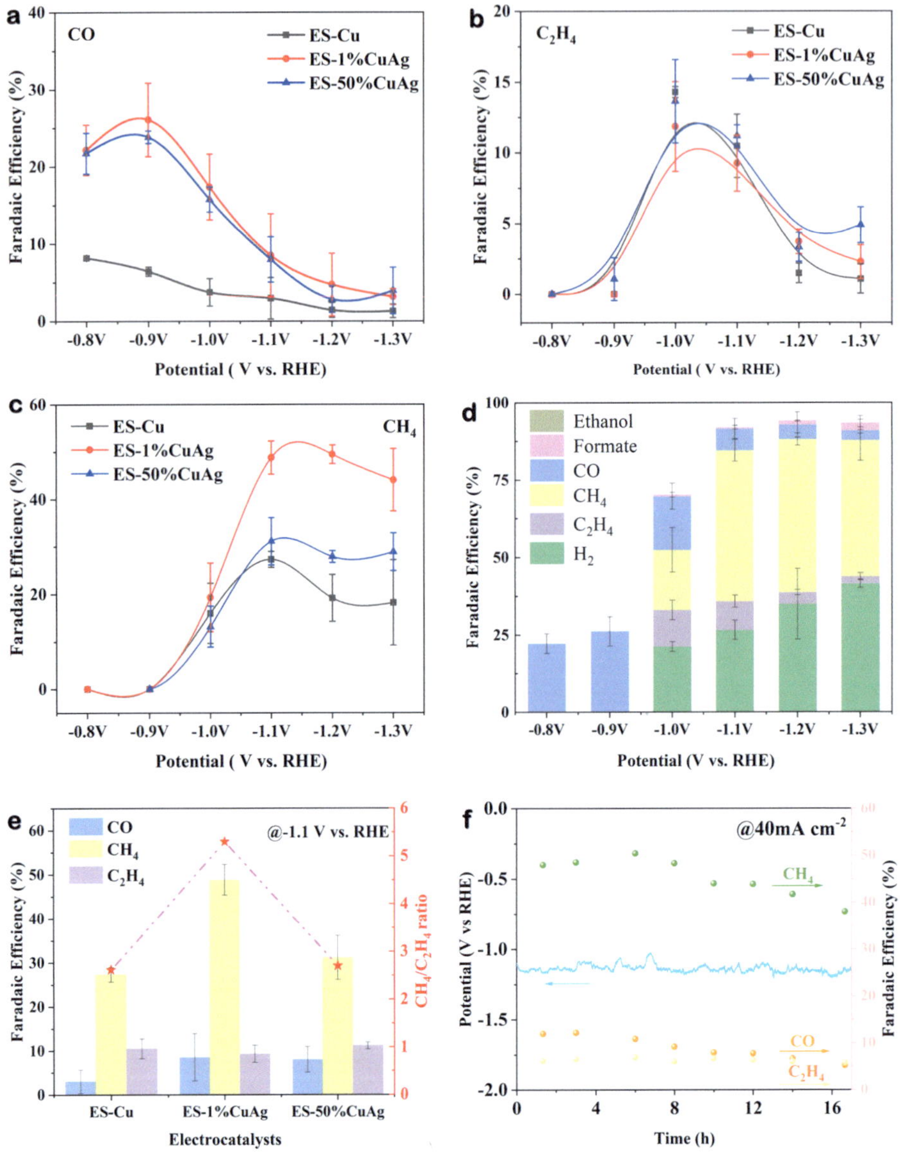

Fig. 3.5 Comparison of Faradaic efficiency of **a** CO, **b** C_2H_4, and **c** CH_4 for ES-Cu, ES-1%CuAg and ES-50%CuAg samples. **d** Faradaic efficiency of reduced products on ES-1%CuAg catalyst as a function of the electrode potential. **e** Comparison of Faradaic efficiency of CH_4, CO and C_2H_4 for ES-Cu, ES-1%CuAg and ES-50%CuAg under an applied potential of $-1.1\ V_{RHE}$. **f** The long-term stability test for ES-1%CuAg catalyst at current density of 40 mA cm^{-2}

attached onto the surface of activated carbon cloth (Fig. A.18a–f); however, a portion of the nanoparticles had clearly detached and dissolved into the electrolyte. Additionally, the catalyst structure and morphology underwent significant reconstruction (Fig. A.18a–f), all of which contributed to the decline in activity.

3.3.4 DFT Calculations

DFT calculations were performed to gain an atomic-level understanding of the origin of high ECR activity of ES-1%CuAg in comparison with ES-Cu and ES-50%CuAg, where GBs of three catalysts were simulated based on the Σ5(210) tilt GB model (Figs. 3.6a and A.19). The following mechanism was assumed for CH_4 production on three catalysts consistent with the ATR FT-IR analysis on reaction intermediates (Fig. 3.4e):

$$
\begin{aligned}
&* + CO_2(g) + (H^+ + e^-) \rightarrow *COOH && (\Delta G_{COOH});\\
&*COOH + (H^+ + e^-) \rightarrow *CO + H_2O && (\Delta G_{CO});\\
&*CO + (H^+ + e^-) \rightarrow *CHO && (\Delta G_{CHO});\\
&*CHO + (H^+ + e^-) \rightarrow *CH_2O && (\Delta G_{CH2O});\\
&*CH_2O + (H^+ + e^-) \rightarrow *CH_3O && (\Delta G_{CH3O});\\
&*CH_3O + (H^+ + e^-) \rightarrow *O + CH_4(g) && (\Delta G_{O});\\
&*O + (H^+ + e^-) \rightarrow *OH && (\Delta G_{OH});\\
&*OH + (H^+ + e^-) \rightarrow *H_2O && (\Delta G_{H2O}).
\end{aligned}
$$

Figure 3.6b displays the free energy diagram for all three examined model structures. The potential determining step (PDS) is *CO reduction for ES-Cu and ES-50%CuAg with $\Delta G_{CHO} = 0.53$ eV and 0.54 eV, respectively, and is *CO_2 reduction to form *COOH for ES-1%CuAg with $\Delta G_{COOH} = 0.45$ eV. We suggest that this shift in PDS is responsible for the better activity of ES-1%CuAg. The adsorption free energy of *CO is a useful descriptor for understanding the differences in activity among the three structures, owing to the scaling relations between *CO and other ECR intermediates. We analyzed the Bader charges to gain a more detailed insight into the electronic structures of active Cu sites. Compared to the active Cu site in ES-Cu, the Bader charge for ES-50%CuAg increases from $+ 0.02|e|$ to $+ 0.05|e|$, which aligns with the observed valence variation (Fig. 3.6c). For ES-1%CuAg, weaker *CO adsorption can be attributed to the formation of Cu–O bonds induced by oxidation, resulting in a Bader charge of $+ 0.25|e|$ for the active Cu site. The weaker *CO adsorption on ES-1%CuAg can be qualitatively illustrated by the fewest electron accumulation around the adsorbed *CO in the charge density difference plots (Fig. 3.6c) and quantitatively supported by the Bader charge analysis that 0.17, 0.12, and 0.21*e* transfer into *CO from ES-Cu, ES-1%CuAg, and ES-50%CuAg, respectively. Owing to the scaling relations, weaker *CO adsorption on ES-1%CuAg results in weaker *CHO and *COOH adsorption leading to $\Delta G_{CHO} = 0.40$ eV and $\Delta G_{COOH} = 0.45$ eV

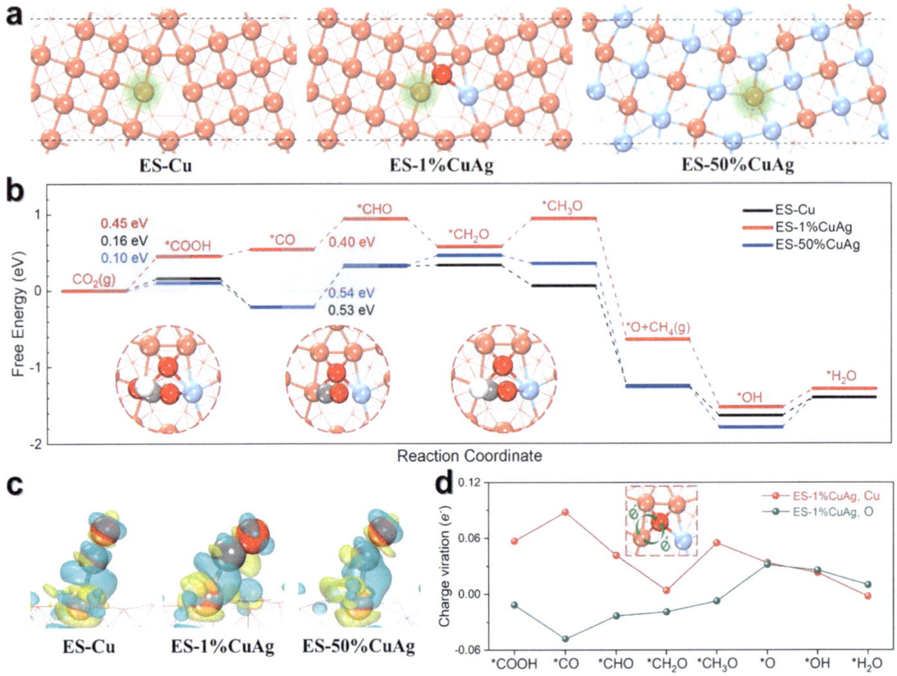

Fig. 3.6 **a** Top views of DFT model structures, where active Cu sites are highlighted in green. **b** Free energy diagram of ECR on ES-Cu, ES-1%CuAg, and ES-50%CuAg at the applied potential of 0 V, including *COOH, *CO and *CHO on ES-1%CuAg. **c** Charge density difference plots of *CO with an isovalue of 0.001 $e/\text{Å}^{-3}$ on ES-Cu, ES-1%CuAg, and ES-50%CuAg, where cyan and yellow regions denote electron accumulation and depletion, respectively. **d** Bader charge variations of active Cu site and O atom in ES-1%CuAg relative to the clean surface during ECR. Color code: Cu, coral; Ag, light blue; C, black; O, red; H, white

(Fig. 3.6b). As a result, the PDS shifts from *CO → *CHO to CO_2 → *COOH. These results explain the role of GBO in ES-1%CuAg in enhancing ECR to CH_4. Moreover, the active $Cu^{\delta+}$ species in ES-1%CuAg has the smallest valence variation during ECR (Fig. A.20), suggesting the highly efficient valence regulation caused by nearby atoms in ES-1%CuAg. Interestingly, for ES-1%CuAg, the trend of valence variation of the O atom is nearly reverse to the active $Cu^{\delta+}$ species, further confirming that the electron transfer mechanism in ES-1%CuAg is via the Cu–O bond, rather than Cu–Ag bond in ES-50%CuAg during ECR (Fig. 3.6d).

3.4 Conclusion

In conclusion, a novel electro-shock method was developed for precise control of catalysts with GBs and GBO structures. A 1% Ag loading in Cu matrix leads to the Ag enrichment on the GBs, triggering the oxidation of Cu sites (i.e., GBO effect). Performance comparisons among the nanoalloy, GBs, and GBO effect highlight superiority of the GBO structure. In situ XAS analysis reveals that, unlike nanoalloy or GBs effect, the performance improvement originates from the steady-state $Cu^{\delta+}$ species and a unique electron transfer mechanism induced by the GBO effect. In situ ATR FT-IR spectra further uncover key intermediate *CHO and *CH_3O in CO_2-to-CH_4 conversion. DFT calculations the further confirm the weaker *CO adsorption on ES-1%CuAg and highly efficient CH_4 conversion pathways. Overall, this work not only establishes the relationship between GBO effect and CO_2-to-CH_4 conversion in bimetallic systems but also provides precise elucidation of active sites derived from CuAg system.

Competing Interests The authors have no conflicts of interests to declare that are relevant to the content of this chapter.

References

1. Herzog, A., et al.: Operando investigation of Ag-decorated Cu_2O nanocube catalysts with enhanced CO_2 electroreduction toward liquid products. Angew. Chem. Int. Ed. **60**(13), 7426–7435 (2021)
2. Nitopi, S., et al.: Progress and perspectives of electrochemical CO_2 reduction on copper in aqueous electrolyte. Chem. Rev. **119**(12), 7610–7672 (2019)
3. Sánchez, O.G., et al.: Recent advances in industrial CO_2 electroreduction. Curr. Opin. Green Sustain. Chem. **16**, 47–56 (2019)
4. Fan, L., et al.: Strategies in catalysts and electrolyzer design for electrochemical CO_2 reduction toward C_{2+} products. Sci. Adv. **6**(8), eaay3111 (2020)
5. Hori, Y.I.: Electrochemical CO_2 reduction on metal electrodes. In: Modern Aspects of Electrochemistry, pp. 89–189 (2008)
6. Hoang, T.T., et al.: Nanoporous copper–silver alloys by additive-controlled electrodeposition for the selective electroreduction of CO_2 to ethylene and ethanol. J. Am. Chem. Soc. **140**(17), 5791–5797 (2018)
7. Chang, Z., et al.: The tunable and highly selective reduction products on Ag@ Cu bimetallic catalysts toward CO_2 electrochemical reduction reaction. J. Phys. Chem. C **121**(21), 11368–11379 (2017)
8. Lei, Q., et al.: Investigating the origin of enhanced C_{2+} selectivity in oxide-/hydroxide-derived copper electrodes during CO_2 electroreduction. J. Am. Chem. Soc. **142**(9), 4213–4222 (2020)
9. Lee, S.Y., et al.: Mixed copper states in anodized Cu electrocatalyst for stable and selective ethylene production from CO_2 reduction. J. Am. Chem. Soc. **140**(28), 8681–8689 (2018)
10. Chen, Z., et al.: Grain-boundary-rich copper for efficient solar-driven electrochemical CO_2 reduction to ethylene and ethanol. J. Am. Chem. Soc. **142**(15), 6878–6883 (2020)

11. Wang, J., et al.: Grain-boundary-engineered La_2CuO_4 perovskite nanobamboos for efficient CO_2 reduction reaction. Nano Lett. **21**(2), 980–987 (2021)
12. Loiudice, A., et al.: Tailoring copper nanocrystals towards C_2 products in electrochemical CO_2 reduction. Angew. Chem. Int. Ed. **55**(19), 5789–5792 (2016)
13. Ma, M., et al.: Controllable hydrocarbon formation from the electrochemical reduction of CO_2 over Cu nanowire arrays. Angew. Chem. Int. Ed. **55**(23), 6680–6684 (2016)
14. Ma, S., et al.: Electroreduction of carbon dioxide to hydrocarbons using bimetallic Cu–Pd catalysts with different mixing patterns. J. Am. Chem. Soc. **139**(1), 47–50 (2017)
15. Lei, Q., et al.: Structural evolution and strain generation of derived-Cu catalysts during CO_2 electroreduction. Nat. Commun. **13**(1), 4857 (2022)
16. Wu, Z., et al.: Grain boundary and interface interaction Co-regulation promotes SnO_2 quantum dots for efficient CO_2 reduction. Chem. Eng. J. **451**, 138477 (2023)
17. Mariano, R.G., et al.: Selective increase in CO_2 electroreduction activity at grain-boundary surface terminations. Science **358**(6367), 1187–1192 (2017)
18. Feng, X., et al.: A direct grain-boundary-activity correlation for CO electroreduction on Cu nanoparticles. ACS Cent. Sci. **2**(3), 169–174 (2016)
19. Van Muylder, J.: Electrochemical Materials Science, Chap. 1. Springer US, Boston, MA (1981)
20. Zhang, W., et al.: Atypical oxygen-bearing copper boosts ethylene selectivity toward electrocatalytic CO_2 reduction. J. Am. Chem. Soc. **142**(26), 11417–11427 (2020)
21. Zhu, Q., et al.: Carbon dioxide electroreduction to C_2 products over copper-cuprous oxide derived from electrosynthesized copper complex. Nat. Commun. **10**(1), 3851 (2019)
22. Mistry, H., et al.: Highly selective plasma-activated copper catalysts for carbon dioxide reduction to ethylene. Nat. Commun. **7**(1), 1–9 (2016)
23. Lee, S.H., et al.: Oxidation state and surface reconstruction of Cu under CO_2 reduction conditions from in situ X-ray characterization. J. Am. Chem. Soc. **143**(2), 588–592 (2020)
24. Yang, Y., et al.: Operando studies reveal active Cu nanograins for CO_2 electroreduction. Nature **614**(7947), 262–269 (2023)
25. Li, C.W., et al.: Electroreduction of carbon monoxide to liquid fuel on oxide-derived nanocrystalline copper. Nature **508**(7497), 504–507 (2014)
26. Wang, Y., et al.: Defect and interface engineering for aqueous electrocatalytic CO_2 reduction. Joule. **2**(12), 2551–2582 (2018)
27. Zhdanov, V.P.: Simulations of oxidation of metal nanoparticles with a grain boundary inside. React. Kinet. Mech. Catal. **130**(2), 685–697 (2020)
28. Dong, W.J., et al.: Grain boundary engineering of Cu–Ag thin-film catalysts for selective (photo) electrochemical CO_2 reduction to CO and CH_4. ACS Appl. Mater. Interfaces **13**(16), 18905–18913 (2021)
29. Xiao, H., et al.: Cu metal embedded in oxidized matrix catalyst to promote CO_2 activation and CO dimerization for electrochemical reduction of CO_2. Proc. Natl. Acad. Sci. U.S.A. **114**(26), 6685–6688 (2017)
30. Lee, C.W., et al.: Defining a materials database for the design of copper binary alloy catalysts for electrochemical CO_2 conversion. Adv. Mater. **30**(42), 1704717 (2018)
31. Stoll, S., Schweiger, A.: EasySpin, a comprehensive software package for spectral simulation and analysis in EPR. J. Magn. Reson. **178**(1), 42–55 (2006)
32. Blöchl, P.E.: Projector augmented-wave method. Phys. Rev. B **50**(24), 17953 (1994)
33. Perdew, J.P., et al.: Generalized gradient approximation made simple. Phys. Rev. Lett. **77**(18), 3865 (1996)
34. Grimme, S.: Semiempirical GGA-type density functional constructed with a long-range dispersion correction. J. Comput. Chem. **27**(15), 1787–1799 (2006)

35. Nørskov, J.K., et al.: Origin of the overpotential for oxygen reduction at a fuel-cell cathode. J. Phys. Chem. B **108**(46), 17886–17892 (2004)
36. Yang, C., et al.: Overcoming immiscibility toward bimetallic catalyst library. Sci. Adv. **6**(17), eaaz6844 (2020)
37. Dong, W.J., et al.: Tailoring electronic structure of bifunctional Cu/Ag layered electrocatalysts for selective CO_2 reduction to CO and CH_4. Nano Energy **78**, 105168 (2020)
38. Mu, S., et al.: Hydroxyl radicals dominate reoxidation of oxide-derived Cu in electrochemical CO_2 reduction. Nat. Commun. **13**(1), 3694 (2022)
39. Lv, X., et al.: Electron-deficient Cu sites on Cu_3Ag_1 catalyst promoting CO_2 electroreduction to alcohols. Adv. Energy Mater. **10**(37), 2001987 (2020)
40. Yi, J.D., et al.: Highly selective CO_2 electroreduction to CH_4 by in situ generated Cu_2O single-type sites on a conductive MOF: stabilizing key intermediates with hydrogen bonding. Angew. Chem. Int. Ed. **59**(52), 23641–23648 (2020)
41. Ma, X.-Y., et al.: Electrolyte-layer-tunable ATR-SEIRAS for simultaneous detection of adsorbed and dissolved species in electrochemistry. Anal. Chem. **94**(32), 11337–11344 (2022)
42. Zhang, Y., et al.: In situ carbon-encapsulated copper-doped cerium oxide derived from MOFs for boosting CO_2-to-CH_4 electro-conversion. ACS Catal. **13**, 1545–1553 (2023)
43. Li, X., et al.: Selective visible-light-driven photocatalytic CO_2 reduction to CH_4 mediated by atomically thin CuIn5S8 layers. Nat. Energy **4**(8), 690–699 (2019)
44. Zijlstra, B., et al.: First-principles microkinetics simulations of electrochemical reduction of CO_2 over Cu catalysts. Electrochim. Acta **335**, 135665 (2020)
45. Zhang, H., et al.: Computational and experimental demonstrations of one-pot tandem catalysis for electrochemical carbon dioxide reduction to methane. Nat. Commun. **10**(1), 3340 (2019)

4 Revealing Real Active Sites in Intricate Grain Boundaries Assemblies on Electroreduction of CO_2 to C_{2+} Products

4.1 Introduction

Electrocatalytic CO_2 reduction (ECR) into value-added chemicals and fuels using renewable energy contributes to global decarbonization, offering an elegant solution for achieving carbon neutrality and fostering sustainable development of human society [1, 2]. However, in the CO_2 activation and conversion process, the rational design of the catalyst plays a central role [2]. To upscale ECR technology, researchers are dedicated to various catalyst design strategies, including grain boundary/interface engineering [3–5], oxidation state modulation [6, 7], alloying [8], and morphology/size effects [9, 10]. Although significant progress has been made in the selectivity of C_1 products (such as CO and formate) [11, 12], the undesired selectivity and activity of C_{2+} products continue to limit the widespread application of ECR. To achieve the highly selective CO_2-to-C_{2+} conversion, more attention should be given to elucidating real-time catalytic mechanisms and establishing the relationship between actual active sites and the catalytic performance of ECR.

Understanding the real active sites and their dynamic evolution of catalyst structures under real reaction conditions is crucial to achieve the aforementioned goals. Under ECR working conditions, factors such as applied potential, electrolyte composition, and CO_2 flow, can all affect the catalyst structure, resulting in phenomena such as atomic rearrangement and changes in chemical states [13, 14]. Therefore, this renders the real active sites of catalysts ambiguous. For instance, oxide-derived copper can form irregular nanosized Cu grains under in situ reduction conditions [6], and the enhanced C_{2+} activity may come from grain boundaries (GBs) or residual oxygen species [15]. At the electrode–electrolyte interface, current crowding phenomena become more pronounced, promoting atomic electromigration and morphology reconstruction [16]. Although high-curvature morphology help expose more active sites, enhancing catalytic activity [9], some studies indicate that

L. Wang et al., *Modulation Strategies of Cu-based Electrocatalysts for Enhancing Electrocatalytic CO_2 Conversion*, Synthesis Lectures on Green Energy and Technology,
https://doi.org/10.1007/978-3-031-98056-5_4

the local electric field of the morphology plays a crucial role [17]. Additionally, it should be noted that the dynamic evolution of active sites may have either beneficial or detrimental effects, resulting in either enhancement or degradation in ECR activity [18]. For example, densely distributed Cu nanoparticles undergo restructuring to form a distorted Cu nanostructure, thereby enhancing the selectivity of multicarbon products [19]. However, detrimental effects may arise from the loss of active sites, and the instability of the catalyst also poses challenges for the industrial application of ECR [20]. Because conventional ex-situ characterizations cannot capture dynamic changes in active sites, the real active sites and dynamic evolution of catalyst structures under actual electrocatalytic reaction conditions remains ambiguous to date.

On the other hand, single strategy often fails to meet practical needs in catalyst design. Coupling multiple structural defects as active sites is often necessary to construct complex catalyst ensembles, where various structural characteristics synergistically interact to significantly enhance catalytic activity [3, 9, 21]. However, identification of the respective promotion or inhibition effects in multiple structural characteristics during ECR has not been thoroughly studied. It remains unclear which design strategy plays a decisive role in the reaction process. Using grain boundaries (GBs), oxidation states, and alloying as examples of ECR performance modulation strategies, it is generally observed that dual-engineering strategies (such as GBs-alloy [22], GBs-oxidation state [23], and oxidation state-alloy [24]) effectively enhance the catalytic performance. For instance, a high density of Cu^0/Cu^+ interfaces derived from Cu_2O–Cu nanocube hybrid catalysts can prevent catalyst reconstruction and enhance the stability of Cu^0/Cu^+ interface structures under high polarization and high current density [23]. However, combining the three performance modulation strategies (GBs, alloying, oxidation states) increases the complexity of investigating their individual contribution. Especially under actual reaction conditions, elucidating their dynamic processes is challenging. For example, compared to antioxidative Cu–Ag films, Cu–Ag films (3nm Ag layer) with GBO degrade the activity and selectivity of ECR towards methane [25]. However, for La_2CuO_4 nanobamboo with abundant twin boundaries, strain at the twin boundary can increase the bond lengths of Cu–O and Cu–La, thereby enhancing ethylene selectivity [26]. Such controversial statements in the literature cannot identify the contribution of each individual design strategy. Therefore, systematic and comprehensive research is needed to reveal the origin of catalytic activity and establish the relationship between each individual design strategies in intricate GBs assemblies and ECR activity.

In this work, GBs are chosen as the starting point, and then oxidation state and alloy phase are introduced to the GBs system. The impact of each modulation strategy in intricate GBs assemblies on ECR performance is systematically investigated through in situ characterizations and activity comparisons. It is noteworthy that water oxidation and low-loading Ag can induce GBO effect, subsequently oxygen from air penetrates into the bulk through GBs, leading to bulk oxidation. In situ X-ray absorption spectroscopy (XAS) spectra reveal the stability of oxidative $Cu^{\delta+}$ species and significant tensile strain of Cu–O and Cu–Cu bonds at GBO sites during ECR. In situ Raman spectroscopy and EPR spectra consistently indicate that stable oxidized $Cu^{\delta+}$ species can be preserved by in situ generated OH radicals, facilitating the CO_2-to-C_{2+} conversion. Theoretical calculations confirm that oxidation of active sites can reduce the energy barrier for the rate-determining step (*CO hydrogenation). Thus, the GBO effect can achieve 70% faradic efficiency for C_{2+} products at $-$ 1.0 V_{RHE} in an H-cell using a neutral electrolyte while it maintains 73.9% faradic efficiency for C_{2+} products at a current density of 150 mA cm^2 in a flow cell using an alkaline electrolyte. This study contributes to bridging the gaps in understanding the synergistic principles of multiple performance modulation strategies in intricate GBs assemblies, highlighting the superiority of GBO and guiding the standardized application of catalyst functional design.

4.2 Experimental and Computational Methods

4.2.1 Materials

Copper(II) nitrate hemi(pentahydrate) ($Cu(NO_3)_2{\cdot}2.5H_2O$), silver nitrate ($AgNO_3$), polyvinylpyrrolidone (PVP) and potassium bicarbonate ($KHCO_3$) were purchased from Sigma-Aldrich, Canada. Potassium hydroxide (KOH) and ultrapure water (18 MΩ cm) were obtained from the University of Waterloo Chemical Store. Carbon paper (Toray HCP-060 CP), Nafion 211 membrane, Fumasep FAB-PK-130 membrane were purchased from fuel cell store, USA. Argon and CO_2 gas were provided by Linde Gas, Canada.

4.2.2 Synthesis of Cu-GB Electrocatalysts

The Cu-GB catalyst is prepared via a simple electrodeposition method [4]. In the typical synthesis process, 1.512g $Cu(NO_3)_2{\cdot}2.5H_2O$ and 0.5 g of PVP were dissolved in 50 mL of ultrapure water, followed by 10 min of ultrasound treatment to ensure homogeneous dispersion. Subsequently, the electrodeposition was conducted using a three-electrode system, with Ag/AgCl and a graphite rod serving as reference and counter electrodes, respectively. A carbon paper with dimensions of 1×1 cm^2 was employed as the working electrode, and subjected to electrodeposition in the above solution at a potential of $-$ 0.3 V (vs. Ag/AgCl) for 30 min. After electrodeposition, the carbon paper was washed with anhydrous ethanol and ultrapure water, and subsequently dried under N_2 and stored under vacuum.

4.2.3 Synthesis of Cu-GBO Electrocatalysts

The Cu-GBO sample was prepared through a straightforward water oxidation treatment, wherein the prepared Cu-GBs sample was immersed in water multiple times, allowing for natural oxidation in air.

4.2.4 Synthesis of Ag/Cu-GBO Electrocatalysts

The introduction of Ag was accomplished via a galvanic replacement reaction. For the Ag/Cu-GBO sample, the Cu-GBs sample was immersed in an ethylene glycol solution containing 0.1 mg/ml $Ag(NO_3)$ at 80 °C for 30 min. The resulting sample was washed with anhydrous ethanol and ultrapure water, then dried under N_2.

4.2.5 Synthesis of CuAg-GBs Electrocatalysts

For the CuAg-GBs sample, the Cu-GBs sample was immersed in a 10 mM $Ag(NO_3)$ aqueous solution for 2 min, followed by washing with anhydrous ethanol and ultrapure water, and dried under N_2.

4.2.6 Ex-Situ Material Characterization

Scanning electron microscopy (SEM, Hitachi S4800, Japan) images were taken with a working accelerating voltage of 10 kV. Glancing-incidence X-ray diffraction (GIXRD) spectra were collected on a MRD diffractometer (PANalytical X'Pert Pro, Netherlands) with Cu Kα radiation (1.54 Å) at an incidence angle of 0.3°. X-ray photoelectron spectroscopy (XPS) measurements were conducted (Thermo-VG Scientific ESCALab 250, USA) using microprobes with a monochromatic Al Kα X-ray source (1486.6 eV). High-angle annular dark-field scanning TEM (HAADF-STEM) and EDS images (FEI Titan 80-300 HB TEM/STEM, USA) were recorded with an aberration corrector operating at 60 kV accelerating voltage.

4.2.7 Electrochemical Measurements

All electrochemical tests were performed using an electrochemical workstation (Gamry Reference 3000, USA) with IR compensation at room temperature. The performance evaluation of electrochemical CO_2 reduction is conducted in both H-cell and flow cell setups.

Initially, a customized H-type cell was assembled, consisting of two compartments separated by a Nafion 211 membrane. An Ag/AgCl reference electrode, Pt counter electrode, and 1 × 1 cm carbon paper is employed as the reference electrode, counter electrode and working electrode, respectively. The 0.1 M $KHCO_3$ electrolyte was saturated by bubbling CO_2 with flow rate of 10 sccm for 30 min before test. For the flow cell testing, a Cu-GBO electrode, Ni foam, and Ag/AgCl are used as the working electrode, counter electrode, and reference electrode, respectively. The gas-tight three-chamber flow cell is equipped with a Fumasep FAB-PK-130 anion exchange membrane (AEM), and 1M KOH serves as the circulating electrolyte with a flow rate of 18 mL/min. The flow rate of CO_2 gas is set to 30 sccm. The ECR catalytic activities were evaluated using the potentiostatic technique for 1000 s. The gas products were analyzed using the Agilent 6890, while the liquid products were characterized using a Bruker 500 MHz Nuclear Magnetic Resonance (NMR) with an internal reference prepared with D_2O and DMSO.

All potentials in this work were measured against the Ag/AgCl reference electrode and converted into reversible hydrogen potential (RHE) using the equation below:

$$E\ (\text{vs. RHE}) = E(\text{vs. Ag/AgCl}) + 0.059 \times \text{pH} + 0.197$$

All the potentials are relative to the RHE unless stated otherwise.

4.2.8 In Situ XAS Measurements

The measurements were conducted at 06ID-1 Hard x-ray MicroAnalysis (HXMA) beamline of the Canadian Light Source (CLS) operated at 2.9 GeV with a constant current of 220 mA. The measurements at the Cu K-edge were performed in fluorescence mode using a Ge detector. The electrolysis was carried out in 0.1M $KHCO_3$ solution with a CO_2 gas flow rate of 30 sccm in a customized cell with thin-layer electrolyte structure and tunable thickness. In situ XAS measurements was recorded at applied cathodic potential from OCP to − 0.8V versus RHE.

4.2.9 In Situ Time-Dependent EPR Measurement

Time-dependent continuous-wave (CW) EPR measurements at X-band frequencies (9.4 GHz) were performed using Bruker EMX Micro EPR spectrometer. DMPO was used as a spin trap. Electrochemistry procedures were performed with Cu-GBO sample in the CO_2-saturated 0.1 M $KHCO_3$ electrolyte containing 10 mM of DMPO. During ECR reaction, 10 uL of samples were taken from the electrochemical cell and transferred into 1 mm ID, 2 mm OD EPR tube (712-SQ-100M Wilmad LabGlass). For the EPR measurements these tubes were placed into 3 mm ID 4 mm OD EPR tubes (707-SQ-250M Wilmad LabGlass). For all the measurements microwave power was set to 1.262 mW (22 dB microwave power attenuation) and modulation amplitude was 2 G at 100 kHz modulation frequency. Simulations of the EPR spectra were performed with EasySpin software [27].

4.2.10 In Situ Raman Measurement

Raman spectra were recorded with a Renishaw inVia Reflex system equipped with a 100× objective and a 633 nm excitation (Renishaw HeNe laser, 17 mW). The filter was set at 5% of maximum intensity. The measurements were performed in a customized gas-diffusion type Raman electrochemical cell with a quartz window. The as-prepared catalysts on carbon paper, Ag/AgCl, and membrane-separated platinum wire were used as the working, reference, and counter electrodes, respectively. The 0.1M $KHCO_3$ electrolyte was constantly purged with CO_2 gas flow rate of 30 sccm.

4.2.11 DFT Calculations

All DFT calculations were spin-polarized and were performed within the Vienna Ab initio Simulation Package (VASP) using the projector augmented wave method, [28] with the generalized gradient approximation (GGA) of Perdew-Burke-Ernzerhof (PBE) [29]. The energy cutoff of the plane-wave basis set was set to be 500 eV, with the Gaussian smearing of the width of 0.05 eV. The convergence criteria were set to be 10^{-6} eV for energy and 0.005 eV for energy difference between two ionic steps. The van der Waals interactions were corrected by the Grimme's DFT-D3 method [30]. For geometrical optimizations and electronic properties calculations, the $3 \times 1 \times 1$ and $9 \times 3 \times 1$ Monkhorst–Pack k-point meshes were respectively used. Cu-GBs were studied by the Σ5(210) tilt GB model, which is generally used for GBs (Fig. B.18a) [31, 32], and then an oxygen atom was added into Cu-GBs to mimic the oxidization effect (Fig. B.18b). For CuAg-GBs, the Cu–Ag Σ5(210) tilt GB model was used (Fig. B.18c, Supporting Information). A planar three-atom Ag structure was put onto Cu-GBs to model Ag cluster, and due to the observation of STEM and EDS images (Figs. B.3a and B.6), an oxygen atom was added (Fig. B.18d). A vacuum space of 15 Å was used for every model to avoid the interlayer interactions, and the bottom one layer was fixed during optimizations. The Gibbs free energy (ΔG) was computed by $\Delta G = \Delta E + \Delta ZPE - T\Delta S$. where ΔE is the reaction energy, T is the temperature, ΔZPE and ΔS are the changes of zero-point energy and entropy, respectively.

4.3 Results

4.3.1 Ex-Situ Characterization of Electrocatalysts

Cu-GBs catalyst, rich in GBs, is prepared via a simple salt electrodeposition method (Fig. 4.1a). Under water oxidation conditions, GBs, as an unstable structural sites, are highly prone to oxidation. Following GBO, more oxygen from air permeates through the GBs into bulk phase, inducing oxidation of active Cu sites from bulk phase and resulting in the formation of the Cu-GBO catalyst. Subsequently, Ag is introduced onto the Cu-GBs sample via a simple galvanic replacement reaction, yielding catalysts with different silver loadings, namely Ag/Cu-GBO and CuAg-GBs catalysts. Grazing incidence X-ray diffraction (GIXRD) was employed to determine the composition and structural characteristics of the prepared samples (Fig. 4.1b). The Cu-GBs catalyst exhibits three distinct XRD diffraction peaks, corresponding to the (111), (200), and (220) reflections of Cu [21]. Under water oxidation and low Ag loading, both Cu-GBO and Ag/Cu-GBO samples show diffraction peaks of Cu_2O [33], indicating the presence of GBO (Fig. 4.1b). The scanning electron microscope (SEM) images show no discernible differences in morphology between the Cu-GBs and Cu-GBO samples (Fig. B.1a, b vs. B.1c, d). Low Ag loading in Ag/Cu-GBs leads to accumulation of Ag at the boundaries (Fig. B.2a, b), whereas high Ag loading in CuAg-GBs causes surface roughening of the initially smooth copper (Fig. B.2c, d).

Aberration-corrected high-angle annular dark-field scanning transmission electron microscopy (AC-HAADF-STEM) was utilized to visualize the structural characteristics of the catalysts. The abundant GBs structures in all catalysts are confirmed (Figs. 4.1c, d and B.3). The corresponding lattice spacing of the Cu(111) and Cu_2O(111) facets are 0.21 nm and 0.25 nm [34, 35] (Fig. 4.1c, d), respectively. Low-loading Ag in Ag/Cu-GBO results in a slight increase in the lattice spacing of Cu (0.22 nm) [36] (Fig. B.3a), while the lattice spacing of the Ag(111) is 0.24 nm [37] (Fig. B.3b, c). Energy-dispersive X-ray spectroscopy (EDS) images further confirm the absence of oxygen in Cu-GBs, showing metallic Cu features (Fig. 4.1e), which is consistent with the GIXRD results (Fig. 4.1b). However, for the Cu-GBO, the electron energy loss spectroscopy (EELS) maps indicate oxygen enrichment at GBs, confirming the existence of GBO (Fig. 4.1f). Subsequently, exposure to air leads to oxygen infiltration (Fig. B.4), inducing bulk oxidation. EELS is a more sensitive technique than EDS and GIXRD. The EELS maps display some weak homogenously oxygen signals in Cu-GBs, which may originate from the adsorbed O_2 or background signals in STEM column during the post modem processes (Fig. B.5). Under low Ag loading, the enrichment of Ag at GBs in Ag/Cu-GBO initially allows oxygen from the air to permeate into the GBs through Ag, inducing GBO and subsequently leading to bulk oxidation (Fig. B.6). With further increases in the Ag content, Ag completely replaces some copper sites, forming GBs structure between Cu and Ag (CuAg-GBs), yet the CuAg-GBs remain unoxidized (oxygen signal is from the

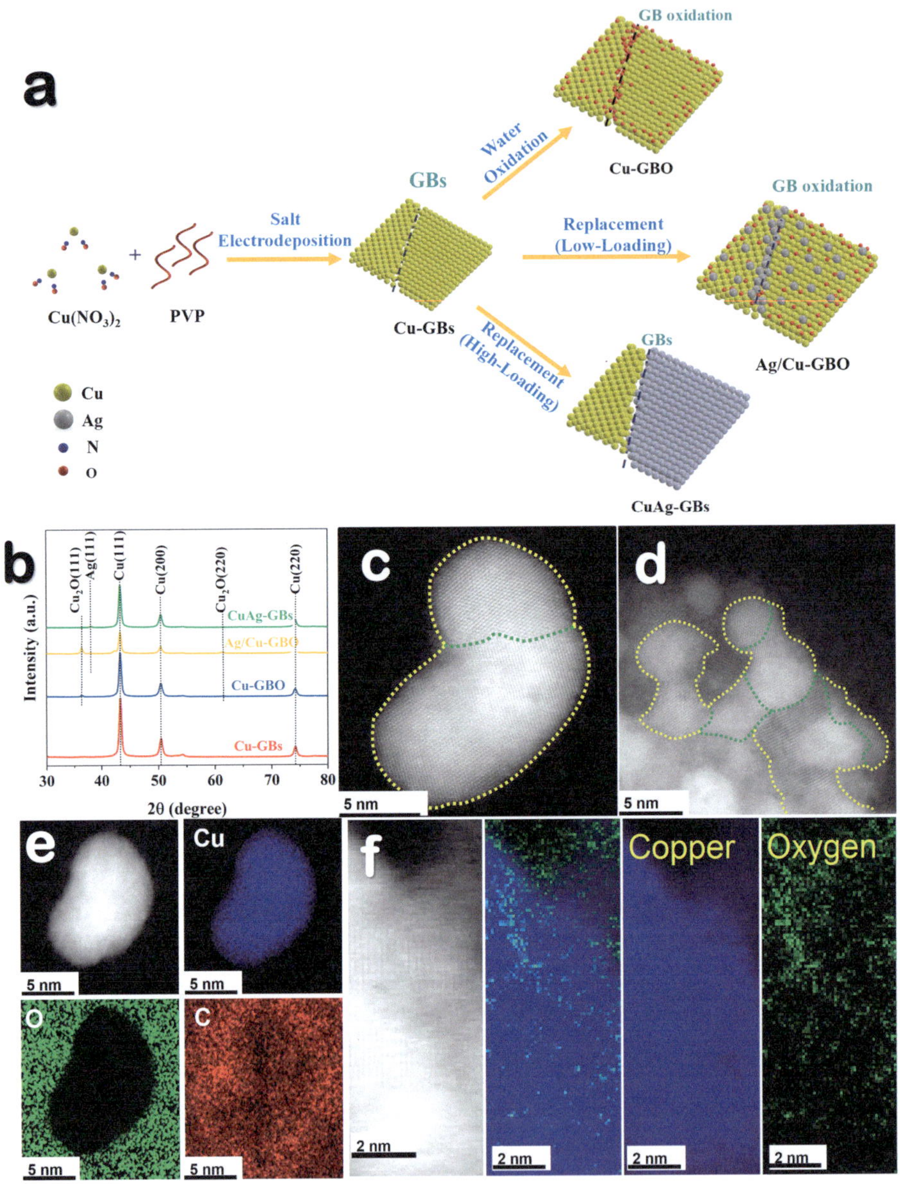

Fig. 4.1 **a** Scheme of the synthesis of electrocatalysts. **b** GIXRD spectra of as-prepared electrocatalysts; **c** AC-HAADF-STEM images of Cu-GBs; **d** AC-HAADF-STEM images of Cu-GBO catalysts; **e** STEM-EDS element mapping of Cu-GBs with the corresponding HAADF-STEM image; **f** EELS maps of Cu-GBO catalysts with the corresponding HAADF-STEM image

background signals), exhibiting the properties of phase-separated CuAg heterostructure (Fig. B.7a, b). This indicates that the oxygen and Ag enrichment at GBs causes the GBO effect, while excessive Ag combined with unsaturated copper sites enhances the resistance of GBs to oxidation.

X-ray photoelectron spectroscopy (XPS) was employed to analyze the surface chemical composition and oxidation states of the catalysts. The Cu $2p_{3/2}$ spectra of all samples display two peaks, corresponding to Cu^{2+} and $Cu^0 + Cu^+$ state [38] (Fig. B.8a), respectively. Among all samples, the Cu-GBs sample exhibits the highest $Cu^{2+}/(Cu^0 + Cu^+)$ intensity ratio (8.28), indicating the facile oxidation of the pristine copper surface when exposed to air during the transfer process for ex situ characterization. However, after GBO treatment, Cu-GBO and Ag/Cu-GBO samples show a stronger $Cu^0 + Cu^+$ signal than Cu^{2+} (Fig. B.8a), attributed to the increased Cu_2O content (Fig. 4.1b). The GBO effect enhances the structural stability of Cu-GBO and Ag/Cu-GBO catalysts, preventing further oxidation of surface Cu_2O. To further clarify the real oxidation state of surface copper sites, Cu LMM Auger spectra was utilized. In the peak fitting of the Cu LMM Auger spectrum (Fig. 4.2a), there are three peaks located at 568.3 eV (metallic Cu and CuO, purple line), 570 eV (Cu_2O, yellow line), and 573.1 eV (representing the transition state of the Cu LMM spectrum, green line) [39]. In terms of the $Cu^+/(Cu^0 + Cu^{2+})$ intensity ratio from Cu LMM spectra, the Cu_2O content is significantly higher in Cu-GBO (1.1) and Ag/Cu-GBO samples (1.0) compared to Cu-GBs (0.4) and CuAg-GBs (0.3). This further demonstrates that the higher $Cu^0 + Cu^+$ signal intensity in the Cu 2p XPS spectrum of Cu-GBO and Ag/Cu-GBO samples is due to the high Cu_2O content, consistent with XPS and GIXRD results. Additionally, the O 1 s XPS spectra can be fitted into three peaks: lattice oxygen species (O_L) at 530.4 eV, surface adsorbed oxygen (O_{ad}) at 531.6 eV, and physically adsorbed water (H_2O) at 533.3 eV [40, 41] (Fig. B.8b). Compared to Cu-GBs and CuAg-GBs samples, the unique presence of O_L in Cu-GBO and Ag/Cu-GBO catalysts further confirms oxygen penetration through GBs, inducing bulk oxidation (Fig. B.8b). In contrast, the Cu_2O phase in Cu-GBs and CuAg-GBs samples originates from the oxidation of O_{ad} species (Fig. B.8b). Moreover, similar Ag 1 s XPS peak positions indicate that Ag/Cu-GBO and CuAg-GBs exhibit similar Ag surface characteristics in the GBs system (Fig. B.8c). The analogous C 1 s XPS spectra for all samples demonstrate the nature of the carbon support, which does not impact the catalyst synthesis [33] (Fig. B.8d).

Synchrotron-based X-ray absorption Near-Edge Structure (XANES) spectra and Extended X-ray Absorption Fine Structure (EXAFS) spectra was utilized to analyze the chemical state and local coordination environment of copper sites. Compared to Cu foil and Cu_2O references, the Cu K-edge XANES spectra indicate that Cu sites in all samples are in oxidized states between 0 and + 1 valence (Fig. 4.2b). The oxidation of Cu sites in Cu-GB and CuAg-GBs may be attributed to spontaneous oxidation of surface Cu atoms in air. However, following GBO induced by water oxidation and low-loading Ag, the absorption edge of copper in the Cu-GBO and Ag/Cu-GBO sample shifts more strongly

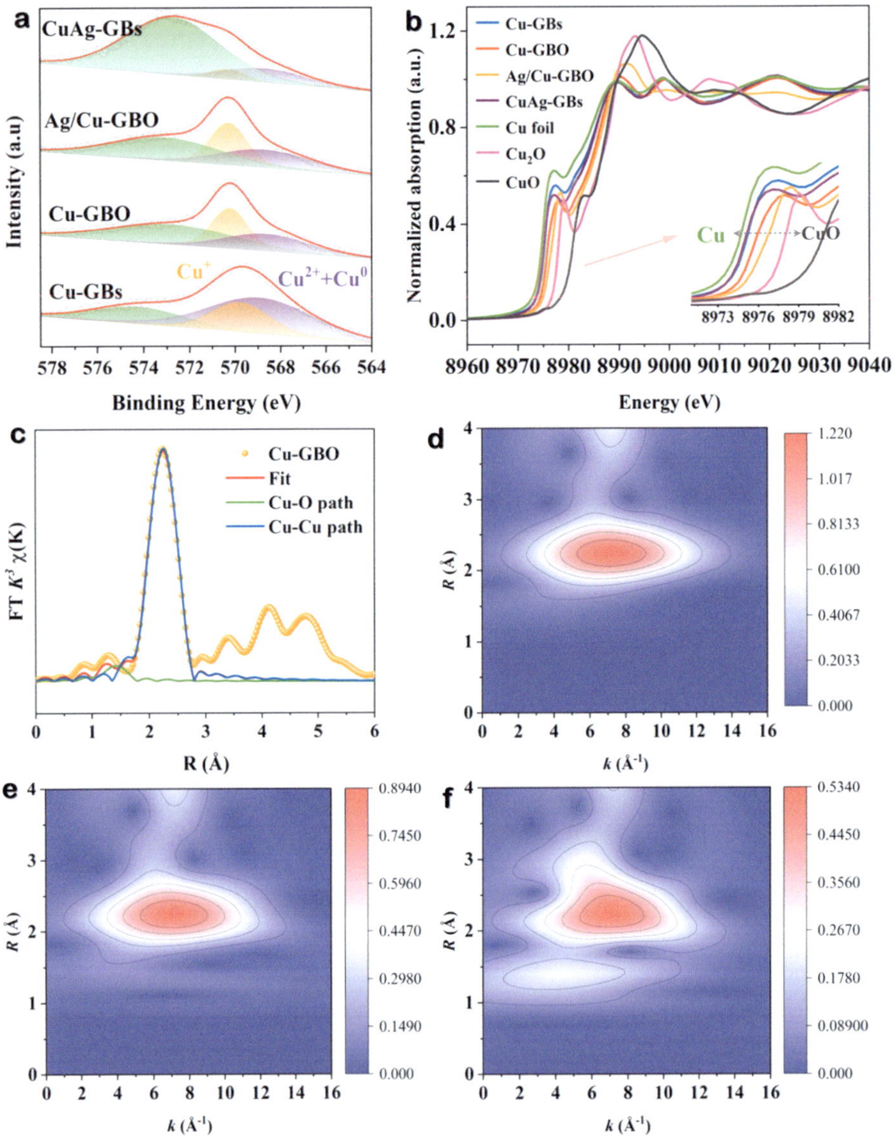

Fig. 4.2 **a** Cu LMM Auger spectra of as-prepared catalysts. **b** Cu K-edge XANES spectra of as-synthesized samples and corresponding references (Cu foil, Cu_2O, and CuO). **c** Cu K-edge EXAFS experimental and fitting spectrum of Cu-GBO catalyst. Wavelet transforms for the k^3-weighted EXAFS signals of Cu foil (**d**), Cu-GBO (**e**), and Ag/Cu-GBO (**f**)

towards higher energy, confirming the presence of GBO effect (Fig. 4.2b). Notably, compared to Cu-GBO, the higher oxidation state in Ag/Cu-GBO samples indicates that Ag doping alters the local electronic structure of copper at the GBs, with excessive oxygen penetrating the GBs exacerbating copper oxidation (Fig. 4.2b). Additionally, fitting in the R-space of EXAFS spectra were conducted to further quantify changes in scattering paths (Figs. 4.2c, B.9, B.10, and Table B.1). Compared to Cu foil, Cu sites in all samples rich in GBs structures are in a low-coordination state. Regardless of water oxidation or Ag doping treatment, there is no significant change in the Cu–Cu bond distance (~ 2.54 Å) for all samples. Nevertheless, after GBO, Cu–O scattering paths can be observed in the Cu-GBO (Fig. 4.2c) and Ag/Cu-GBO (Fig. B.9d) samples. The different Cu–O coordination numbers (0.4 at 1.81 Å for Cu-GBO and 1.1 at 1.84 Å for Ag/Cu-GBO) suggest that compared to a water oxidation-induced GBO effect, oxygen is more likely to penetrate the GBs and bind with unsaturated Cu sites at the GBs facilitated by Ag enrichment. Wavelet transform was employed to elucidate the coordination environment near Cu sites and confirm the atomic distribution of Cu in all samples. A strong wavelet transform signal centered at 7.2 $Å^{-1}$, originating from Cu–Cu coordination, is observed in Cu foil (Fig. 4.2d), while two other intense signals at 5.0 and 4.6 $Å^{-1}$ for CuO (Fig. B.11a) and Cu_2O (Fig. B.11b) references could be attributed to Cu–O and Cu–Cu coordination, respectively. The wavelet transform contour plots for Cu-GBs (Fig. B.11c) and CuAg-GBs (Fig. B.11d) only exhibit Cu–Cu coordination characteristics. For samples with GBO, first shell Cu–O and second shell Cu–Cu scattering distributions indicate that GBO effect leads to changes in the local coordination environment around Cu (Fig. 4.2e, f).

4.3.2 In Situ Multimodal Characterization of Electrocatalysts for ECR in Neutral Media

In situ X-ray absorption spectroscopy was employed to overcome the limitations of ex-situ characterization techniques. This in situ characterization reveals the intrinsic active sites and their dynamic evolution during the ECR reaction process. The Cu K-edge XANES and EXAFS spectra were recorded under various ECR reaction conditions including open circuit potential (OCP), OCP after cyclic voltammetry (CV), applied cathodic potentials (0 V_{RHE} and $-$ 0.8 V_{RHE}), and OCP after reaction (Figs. 4.3a, b and B.12a–d). For Cu-GBOs samples, in situ Cu K-edge XANES spectra indicate a slight reduction of $Cu^{\delta+}$ species ($0 < \delta < 1$) during ECR, while the oxidized $Cu^{\delta+}$ species still persist under reducing potentials (Fig. 4.3a), demonstrating their structural stability induced by the GBO effect. In the corresponding in situ Cu K-edge EXAFS spectra, the Cu–Cu bond length changes by 0.07 Å between OCP before and after the reaction (Fig. 4.3b). The Cu–O bond length increases by 0.15 Å after the ECR reaction (Fig. 4.3b). These notable changes indicate irreversible reconstruction of the catalyst's active sites, such as local environment of Cu sites, during the ECR reaction (Fig. 4.3b). Notably, after applying the low cathodic potential (0 V_{RHE}), the Cu–O bond length increase by 0.05 Å and further increases with higher reducing potentials (− 0.8 V_{RHE}) (Fig. 4.3b), indicating that the enhancement in ECR activity is closely related to the intensity of tensile strain at the active sites. The strong tensile strain of the Cu–O bond can significantly promote the formation of C_{2+} products. Even upon cessation and switching of the applied potential to OCP, the length of the Cu–O bond persists in its increase (Fig. 4.3b), further indicating that the active $Cu^{\delta+}$ sites show an irreversible and strong tensile strain effect.

In contrast, a slight reduction of $Cu^{\delta+}$ species during the ECR process is also observable in Cu-GBs (Fig. B.12a) and Ag/Cu-GBOs (Fig. B.12c). The $Cu^{\delta+}$ species does not fully reduce to metallic Cu, even at high cathodic potentials, confirming that both unoxidized and oxidized GBs can strengthen the structural stability and resistance to the

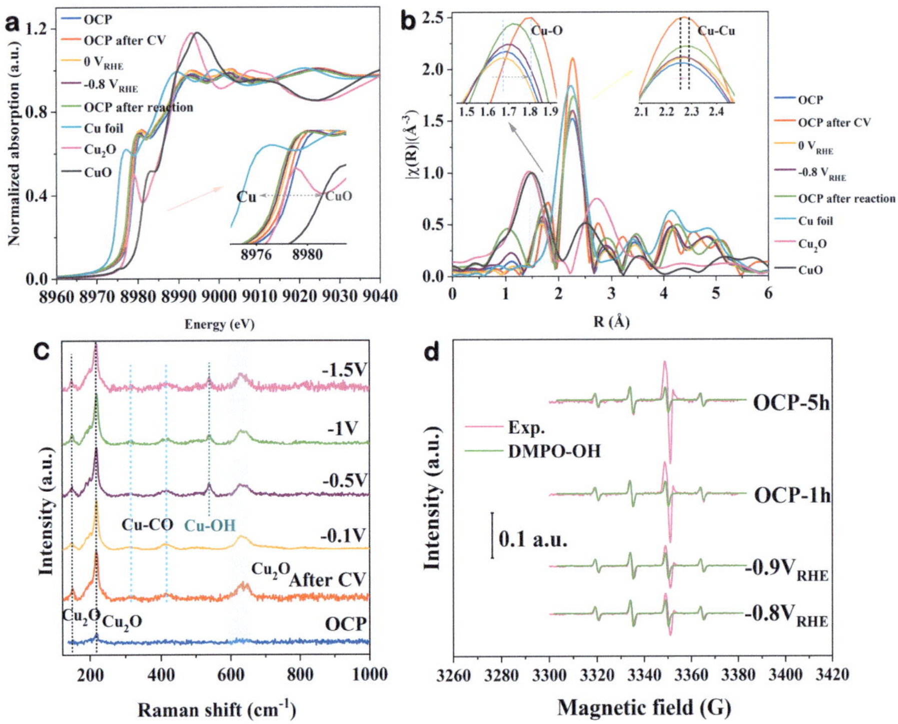

Fig. 4.3 **a** In situ Cu K-edge XANES spectra and **b** in situ Cu K-edge EXAFS spectra of Cu-GBO at different applied potentials (including open circuit potential (OCP), OCP after cyclic voltammetry (CV), applied cathodic potentials (0 V_{RHE} and − 0.8 V_{RHE}), and OCP after reaction); **c** In situ Raman spectra of Cu-GBO catalyst at different applied potentials (− 0.1 V, − 0.5 V, − 1 V, and − 1.5 V). **d** In situ time-dependent normalized EPR spectra of Cu-GBO sample at − 0.8 V and − 0.9 V versus RHE, followed by switching to OCP for different periods on Cu-GBO sample. Experimental spectra are depicted in pink while simulated spectra are depicted in dark green for DMPO-OH radical, respectively

electroreduction during ECR. However, the in situ Cu K-edge EXAFS spectra show stable Cu–Cu bonds for Cu-GBs (Fig. B.12b) and Ag/Cu-GBO (Fig. B.12d) during the ECR reaction, unlike Cu-GBO (Fig. 4.3b). It is noteworthy that, the absence of Cu–O bonds in the ex-situ EXAFS spectra of the original Cu-GBs samples, but the appearance of Cu–O bond during ECR reactions, suggests that some active Cu species undergo oxidation. This could be due to the in situ generated strongly oxidative species, causing spontaneous oxidation of unstable copper species in the GBs system [42–44] (Fig. B.9b vs. B.12b). Additionally, the Cu–O bonds in Cu-GBs exhibit a tensile strain effect similar to that in Cu-GBOs during the ECR reaction (Fig. B.12b). In contrast to the irreversible strain effect after the reaction in Cu-GBO (Fig. 4.3b), the Cu–O bond length reverts to its initial state in Cu-GBs (Fig. B.12b), indicating an elastic tensile strain effect in Cu-GBs. This suggests that the original GBs structure does not lead to the irreversible reconstruction of active sites, which is more favorable for structural stability of active sites compared to the GBO structure. Furthermore, similar alterations in Cu–O bond lengths are evident in Ag/Cu-GBO (changing by 0.07 Å) as in Cu-GBO (Fig. B.12d). This consistency indicates the uniform impact of GBO effects on Cu–O bonds.

In situ Raman spectra were utilized to monitor the structural evolution of Cu-GBO during the ECR reaction. The freshly prepared Cu-GBO sample displays Raman peaks of Cu_2O (149, 217, and 637 cm^{-1}) in CO_2-saturated 0.1 M $KHCO_3$ solution (Fig. 4.3c) [43, 45]. According to the Pourbaix diagram of Cu, oxidized copper species should be fully reduced to metallic copper under cathodic potentials [46]. However, upon applying a cathodic potential of − 0.1 V, the characteristic peaks of Cu_2O not only persist but also new Raman bands attributed to Cu-CO at 302 and 406 cm^{-1} emerge (Fig. 4.3c) [47–49]. Conversely, CO-related vibrations and active oxidative copper species cannot be observed in Cu-GBs (Fig. B.13). This indicates that the GBO structures, distinct from the GBs, effectively promote the activation and conversion of CO_2 and enhance the ECR activity, with oxidative $Cu^{\delta+}$ species serving as the active site during the ECR reaction (Fig. 4.3c). Notably, with the increase in negative potentials, Cu-OH species at 532 cm^{-1} begin to appear and remain stable (Fig. 4.3c) [50]. This suggests that the potential-induced OH radicals enrich the catalyst surface, altering local reaction environment on the catalyst surface and stabilizing $Cu^{\delta+}$ species.

In situ time-dependent electron paramagnetic resonance (EPR) spectra were utilized to probe the presence of highly oxidative hydroxyl (OH) species during the ECR reaction. The Cu-GBO sample undergoes a 30-min ECR electrolysis at cathodic potentials of − 0.8 V_{RHE} and − 0.9 V_{RHE} to mitigate the influence of residual oxidized species in the CO_2-saturated 0.1 M $KHCO_3$ electrolyte using 10 mM 5,5-dimethyl-1-pyrroline N-oxide (DMPO) (Fig. 4.3d). Interestingly, in our system, H radicals are not observed upon applying the potentials, whereas OH radicals (A_N = 1.50 mT, A_H = 1.48 mT) are produced during the ECR reaction. This is different observation with prior literature, where H radicals from the hydrogen evolution reaction (HER) are typically observed [43]. Upon switching the potentials to OCP, the highly oxidative hydroxyl species persist throughout the OCP process (Fig. 4.3d). This reinforces the notion that the in situ generated strongly oxidative hydroxyl species alter local reaction environment, prompting the rapid oxidation of active $Cu^{\delta+}$ sites and consequently stabilizing $Cu^{\delta+}$ species during the ECR.

4.3.3 Performance Evaluation of ECR

To highlight the GBO effect on C_{2+} selectivity, the ECR performance was evaluated in a neutral electrolyte (0.1 M $KHCO_3$) using an H-cell. All catalysts exhibit similar products: CO, formate, and CH_4 as the C_1 products, and ethylene and ethanol as C_{2+} products (Figs. 4.4a, B.14, B.15, and B.16). Cu-GBs predominantly produces CO at low potentials (− 0.8 V_{RHE} and − 0.9 V_{RHE}), but at higher potentials (− 1.0 V_{RHE} and − 1.1 V_{RHE}), the CO selectivity decreases significantly (Fig. B.16a). The potential-induced promotion effect facilitates the transformation of *CO into higher products (C_2H_4, ethanol and CH_4) via proton-coupled electron transfer (PECT), with ethanol produced to a lesser amount (Fig. B.16a). At − 1.1 V_{RHE}, Cu-GBs achieves maximum Faradaic efficiency (FE) with 43.3% for CH_4 and 31.6% for C_2H_4 (Figs. 4.4b and B.16a). This indicates that GBs structures in Gu-GBs exhibit comparable selectivity in the reaction pathways for CH_4 and C_{2+} products. However, following the introduction of GBO, oxidized $Cu^{\delta+}$ species significantly inhibit the reaction pathway of CH_4, resulting in a notable enhancement in the C_{2+} selectivity (Fig. 4.4a, b). Therefore, the Cu-GBO sample shows an exceptional 70% selectivity for C_{2+} products at − 1.0 V_{RHE}, with a C_2H_4 FE of 42% and ethanol FE of 28% (Fig. 4.4a, b). The substantial variance in the catalytic performance of Cu-GBs and Cu-GBO regarding selectivity towards CH_4 and C_{2+} product highlights the significance of oxidized $Cu^{\delta+}$ species as the real active sites in the CO_2-to-C_{2+} conversion (Fig. 4.4b). Furthermore, the onset potentials for ethylene and ethanol, as well as the optimal activity potentials, are lower in Cu-GBO compared to Cu-GBs (Figs. 4.4a vs. B.16a). This indicates that the GBO effect can substantially enhance the intrinsic catalytic activity of GBs. Upon the introduction of Ag into the GBs structure, regardless of the Ag loading, there is a notable increase in the CO FE, while the suppression of the CH_4 reaction pathway is not substantial, leading to decreased selectivity towards C_{2+} products (Figs. 4.4b and B.16b, c). Consequently, Ag/Cu-GBO shows a selectivity of 28.5% FE for ethylene, 32.2% FE for CH_4, and 9.4% FE for ethanol at − 1.1 V_{RHE} (Figs. 4.4b and B.16b), while CuAg-GBs exhibits a selectivity of 26% FE for ethylene, 31% FE for methane, and 12% FE for ethanol at − 1.1 V_{RHE} (Figs. 4.4b and B.16c). Similar electrocatalytic performance suggests that the GBO effect induced by low-loading Ag and the GBs effect induced by high-loading Ag have comparable reaction energy barriers in the ECR conversion to specific products under neutral conditions (Fig. 4.4b). Compared to Cu-GBs, the decreased CH_4 performance of CuAg-GBs highlights the superior activity of original GBs for CH_4 production (Fig. 4.4b). While CuAg-GBs promotes ethanol production, they also reduce ethylene selectivity (Fig. B.16c). Although both Ag/Cu-GBO and

Cu-GBO samples exhibit GBO effect, the notable disparity in inhibiting CH_4 production and achieving high C_{2+} products selectivity suggests that the Ag nanoclusters on the Cu surface (i.e. nanoalloy effect) in Ag/Cu-GBO enhance CO and CH_4 production compared to Cu-GBO, which demonstrates a synergistic effect. However, for the C_{2+} products, the nanoalloy effect competes with the GBO effect. (Fig. 4.4b). The nanoalloy effect in our Ag/Cu-GBO system does not follows the tandem fashion owing to the specific Ag-enrichment GBO structure with low-coordination Cu sites. Ag and Cu lack geometric arrangements to exhibit synergistic tandem catalytic effects; instead, they provide unique adsorption sites distinct from those offered by GBO for active intermediates, thereby enhancing product diversity. This emphasizes the crucial role of active $Cu^{\delta+}$ species in the conversion of CO_2-to-C_{2+} products. The nanoalloy effect in Ag/Cu-GBO, the GBs effect in Cu-GBs, and the GBs effect in CuAg-GBs show a tendency towards the CH_4 conversion in neutral conditions. In contrast, the GBO effect in Cu-GBO enhances the production of C_{2+} products under neutral conditions.

To investigate the impact of the electrolyte on catalytic performance, we employed alkaline (1 M KOH) and neutral (0.1 M $KHCO_3$) electrolyte to assess the ECR activity of

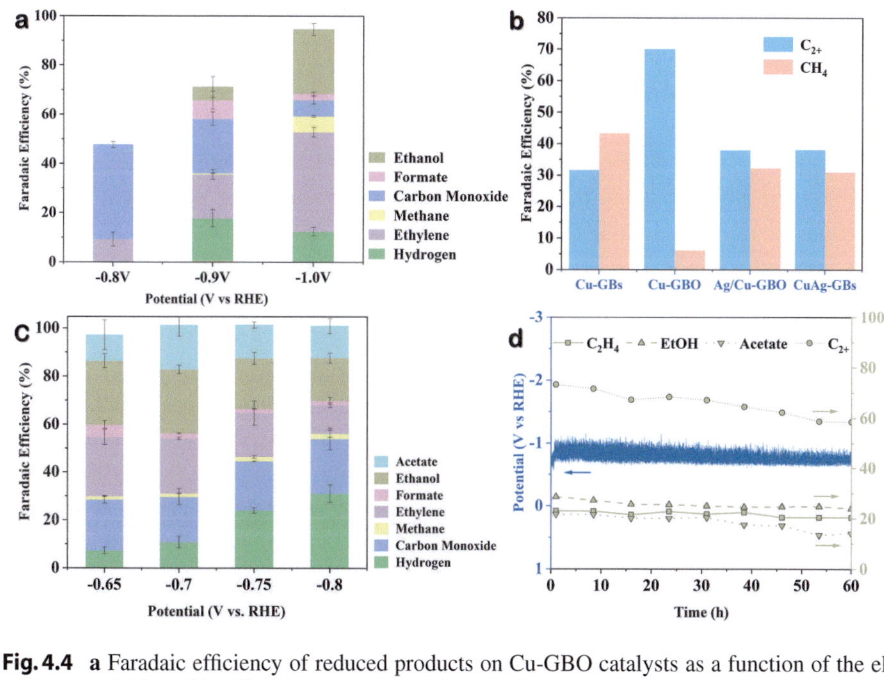

Fig. 4.4 **a** Faradaic efficiency of reduced products on Cu-GBO catalysts as a function of the electrode potential in an H-cell under neutral conditions. **b** The comparison of C_{2+} and CH_4 selectivity among all samples. **c** Faradaic efficiency of reduced products for Cu-GBO under applied potentials in a flow cell under alkaline conditions. **d** Electrochemical stability of Cu-GBO at a current density of 150 mA cm^{-2} in a flow cell under alkaline conditions

Cu-GBO samples in a flow cell. Interestingly, the inhibited acetate production observed in a neutral electrolyte is significantly alleviated in alkaline electrolyte (Figs. 4.4c vs. B.17). At potentials ranging from $-$ 0.65 V_{RHE} to $-$ 0.8 V_{RHE} in alkaline electrolyte, the C_{2+} products such as ethylene, ethanol, and acetate are predominant (Fig. 4.4c). At $-$ 0.7 V_{RHE}, the total FE for C_{2+} products is 73.9% under a current density of ~ 150 mA cm^{-2}, including 23.7% FE for ethylene, 28.6% FE for ethanol, and 21.6% FE for acetate (Fig. 4.4c). Under neutral conditions, the current density (~ 35 mA cm^{-2} at $-$ 0.85 V_{RHE}) and the FE for C_{2+} products (35%) are significantly lower than those under alkaline conditions (Fig. B.17). This indicate that alkaline electrolytes not only increase current density by more than four times, but also facilitate the C–C coupling step, thereby enhancing the conversion of CO_2 towards more favorable products pathways. Long-term stability tests were conducted to evaluate the practical applicability of the catalyst. Considering the selectivity of C_{2+} products and the industrial-relevant current density, Cu-GBO was selected to run stability test in a flow cell with 1 M KOH at a constant current of 150 mA cm^{-2} (Fig. 4.4d). The catalyst could operate continuously for 60 h. Although there is a slight decline in the FE of various C_{2+} products, the overall FE decrease only from the original 73.4% to 58.5%. This indicates the potential application of Cu-GBO in industrial-scale ECR for C_{2+} products.

4.3.4 DFT Calculations

DFT calculations were performed to gain deep insights into the remarkable activity of Cu-GBO, with ethylene being the representative C_{2+} product. For the CO_2-to-ethylene conversion, the rate-determining step (RDS) is always the hydrogenation of *CO to *CHO (Fig. 4.5a), with $\Delta G_{RDS} = 0.71$, 0.81, 1.01, and 1.05 eV for Cu-GBO, Cu-GBs, CuAg-GBs and Ag/Cu-GBO, respectively. The C–C coupling, *CHO → *CHOCO, follows the RDS and is thermodynamically favorable owing to the presence of dual active sites (1st and 2nd active sites, Fig. B.18). For Cu-GBs, Cu-GBO and CuAg-GBs, *CO binds solely to one Cu site (1st Cu sites), while for Ag/Cu-GBO, it bonds to both Cu and Ag sites (Fig. B.19). The electronic structures of the former three models were further investigated to elucidate the dominant factor influencing ECR activity. The charge density difference analysis of *CO reveals the smallest electron accumulation around the carbon atom when *CO is absorbed on Cu-GBO (Figs. 4.5b–d and B.20). This carbon atom is likely the least stabilized, thus requiring the minimum energy for subsequent hydrogenation (i.e., smallest ΔG_{*CHO}). The projected density of states (PDOS, Fig. 4.5e) confirms this, as strong *CO binding deepens the electronic states of the carbon atom, with the p-band center (ε_p) of carbon atom of *CO on Cu-GBO being closest to the Fermi level. The preference of *CHO over *COH (formed by the hydrogenation of the oxygen atom in *CO) is due to the higher ε_p value of the carbon atom compared to the corresponding oxygen atom (Fig. B.21). Figure 4.5f shows that ΔG_{*CHO} decreases with the increasing $E_{ads\text{-}CO}$, further

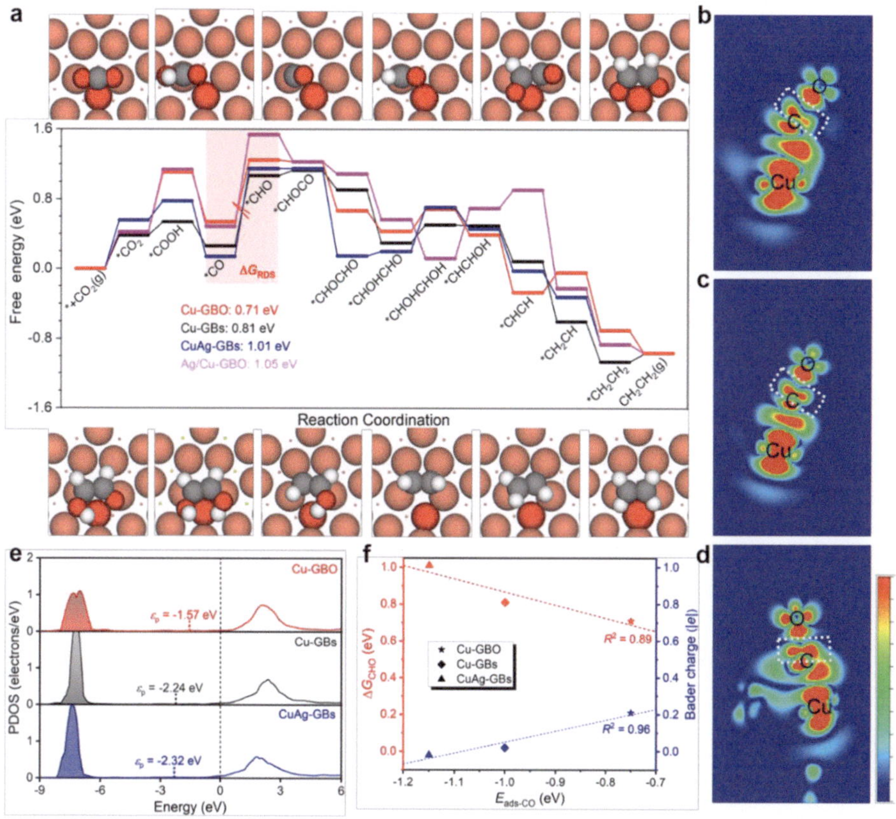

Fig. 4.5 **a** The free energy profile of CO_2-to-ethylene conversion on Cu-GBs, Cu-GBO, Ag/Cu-GBO, and CuAg-GBs at the applied potential of 0 V, including the optimized intermediates on Cu-GBO. Color code: Cu, coral; C, black; O, red; H, white. 2D plots of charge density differences of *CO on **b** Cu-GBO, **c** Cu-GBs, and **d** CuAg-GBs plotted from 0 (blue) to 0.05 e^-/Bohr^3 (red). The white dotted boxes are the main electron accumulations of carbon atoms. **e** PDOS of p orbitals of carbon atoms of *CO on Cu-GBs, Cu-GBs, and CuAg-GBs. **f** ΔG_{*CHO} and Bader charge as functions of $E_{\text{ads-CO}}$ for Cu-GBs, Cu-GBs, and CuAg-GBs, respectively

indicating that weak *CO adsorption facilitates the subsequent hydrogenation to *CHO due to the reduced stabilization of the carbon atom. Typically, a higher d-band center (ε_d) corresponds to stronger adsorption [51]. However, despite Cu-GBO having a higher ε_d value, it exhibits weaker *CO adsorption compared to Cu-GBs, indicating another factor influencing *CO adsorption (Fig. B.22). Additionally, Fig. 4.5f shows a nearly linear relationship between $E_{\text{ads-CO}}$ and the Bader charge value of the 1st Cu active site, suggesting the significance of active site oxidation in *CO adsorption. Consequently, in the CO_2-to-C_2 product conversion, a higher oxidation state of the active Cu sites induces weaker *CO adsorption, resulting in a smaller ΔG_{RDS} value.

4.4 Conclusion

In conclusion, we engineered four different GB structures to elucidate the real active sites and their dynamic evolution processes within complex GBs assembles. In situ XAS spectroscopy reveals that while GBs structures bolster the resilience of active $Cu^{\delta+}$ species against electrochemical reduction, the stabilized Cu-Cu bonds and weak tensile strain of Cu–O bond inadequately augments the selectivity towards C_{2+} products. In situ EPR spectra demonstrate that in situ generated strongly oxidative OH radicals in the neutral electrolyte leads to the spontaneous oxidation of active copper sites. In situ Raman spectroscopy, coupled with performance comparisons, confirms the pivotal role of the oxidation state in CO_2-to-C_{2+} conversion. DFT calculations indicate that the oxidation state reduces the energy barrier of the rate-determining step (*CO hydrogenation), thus promoting C–C coupling. This study clarifies the impact of each performance modulation strategy on catalytic activity, aiding in the avoidance of blindly combining multiple design strategies.

Competing Interests The authors have no conflicts of interests to declare that are relevant to the content of this chapter.

References

1. Resasco, J., Bell, A.T.: Electrocatalytic CO_2 reduction to fuels: progress and opportunities. Trends Chem. **2**(9), 825–836 (2020)
2. Ye, W., et al.: A review on electrochemical synthesized copper-based catalysts for electrochemical reduction of CO_2 to C_{2+} products. Chem. Eng. J. **414**, 128825 (2021)
3. Wu, Z., et al.: Grain boundary and interface interaction Co-regulation promotes SnO_2 quantum dots for efficient CO_2 reduction. Chem. Eng. J. **451**, 138477 (2023)
4. Chen, Z., et al.: Grain-boundary-rich copper for efficient solar-driven electrochemical CO_2 reduction to ethylene and ethanol. J. Am. Chem. Soc. **142**(15), 6878–6883 (2020)
5. Chang, X., et al.: Tuning Cu/Cu_2O interfaces for the reduction of carbon dioxide to methanol in aqueous solutions. Angew. Chem. **130**(47), 15641–15645 (2018)
6. Lei, Q., et al.: Investigating the origin of enhanced C_{2+} selectivity in oxide-/hydroxide-derived copper electrodes during CO_2 electroreduction. J. Am. Chem. Soc. **142**(9), 4213–4222 (2020)
7. Lee, S.Y., et al.: Mixed copper states in anodized Cu electrocatalyst for stable and selective ethylene production from CO_2 reduction. J. Am. Chem. Soc. **140**(28), 8681–8689 (2018)
8. Okatenko, V., et al.: Alloying as a strategy to boost the stability of copper nanocatalysts during the electrochemical CO_2 reduction reaction. J. Am. Chem. Soc. **145**(9), 5370–5383 (2023)
9. De Luna, P., et al.: Catalyst electro-redeposition controls morphology and oxidation state for selective carbon dioxide reduction. Nat. Catal. **1**(2), 103–110 (2018)
10. Zhang, X., et al.: Ultra-small Cu nanoparticles embedded in N-doped carbon arrays for electrocatalytic CO_2 reduction reaction in dimethylformamide. Nano Res. **11**(7), 3678–3690 (2018)
11. Zheng, T., et al.: Large-scale and highly selective CO_2 electrocatalytic reduction on nickel single-atom catalyst. Joule. **3**(1), 265–278 (2019)

12. Li, L., et al.: Stable, active CO_2 reduction to formate via redox-modulated stabilization of active sites. Nat. Commun. **12**(1), 5223 (2021)
13. Liu, L., Corma, A.: Structural transformations of solid electrocatalysts and photocatalysts. Nat. Rev. Chem. **5**(4), 256–276 (2021)
14. Tomboc, G.M., et al.: Potential link between Cu surface and selective CO_2 electroreduction: perspective on future electrocatalyst designs. Adv. Mater. **32**(17), 1908398 (2020)
15. Lei, Q., et al.: Structural evolution and strain generation of derived-Cu catalysts during CO_2 electroreduction. Nat. Commun. **13**(1), 4857 (2022)
16. Li, F., et al.: Interplay of electrochemical and electrical effects induces structural transformations in electrocatalysts. Nat. Catal. **4**(6), 479–487 (2021)
17. Dong, W.J., et al.: Electric-field-driven electrochemical CO_2 reduction of sharpened Sn/Cu catalysts. Appl. Surf. Sci. **565**, 150460 (2021)
18. Lai, W., et al.: Dynamic evolution of active sites in electrocatalytic CO_2 reduction reaction: fundamental understanding and recent progress. Adv. Funct. Mater. **32**(16), 2111193 (2022)
19. Li, Y., et al.: Electrochemically scrambled nanocrystals are catalytically active for CO_2-to-multicarbons. Proc. Natl. Acad. Sci. U.S.A. **117**(17), 9194–9201 (2020)
20. Popović, S., et al.: Stability and degradation mechanisms of copper-based catalysts for electrochemical CO_2 reduction. Angew. Chem. **132**(35), 14844–14854 (2020)
21. Gu, Z., et al.: Oxygen vacancy tuning toward efficient electrocatalytic CO_2 reduction to C_2H_4. Small Methods. **3**(2), 1800449 (2019)
22. Choi, C., et al.: Intimate atomic Cu–Ag interfaces for high CO_2 RR selectivity towards CH_4 at low over potential. Nano Res. **14**, 3497–3501 (2021)
23. Wu, Q., et al.: Nanograin-boundary-abundant Cu_2O-Cu nanocubes with high C_{2+} selectivity and good stability during electrochemical CO_2 reduction at a current density of 500 mA/cm^2. ACS Nano (2023)
24. Patra, K.K., et al.: Boosting electrochemical CO_2 reduction to methane via tuning oxygen vacancy concentration and surface termination on a copper/ceria catalyst. ACS Catal. **12**(17), 10973–10983 (2022)
25. Dong, W.J., et al.: Grain boundary engineering of Cu–Ag thin-film catalysts for selective (photo) electrochemical CO_2 reduction to CO and CH_4. ACS Appl. Mater. Interfaces **13**(16), 18905–18913 (2021)
26. Wang, J., et al.: Grain-boundary-engineered La_2CuO_4 perovskite nanobamboos for efficient CO_2 reduction reaction. Nano Lett. **21**(2), 980–987 (2021)
27. Stoll, S., Schweiger, A.: EasySpin, a comprehensive software package for spectral simulation and analysis in EPR. J. Magn. Reson. **178**(1), 42–55 (2006)
28. Bliichl, P.: Projector augmented-wave method. Phys. Rev. B **50**(24), 17953–17979 (1994)
29. Perdew, J.P., et al.: Generalized gradient approximation made simple. Phys. Rev. Lett. **77**(18), 3865 (1996)
30. Grimme, S.: Semiempirical GGA-type density functional constructed with a long-range dispersion correction. J. Comput. Chem. **27**(15), 1787–1799 (2006)
31. Huang, Z., et al.: Uncovering the influence of common nonmetallic impurities on the stability and strength of a Σ5 (310) grain boundary in Cu. Acta Mater. **148**, 110–122 (2018)
32. Zheng, H., et al.: Grain boundary properties of elemental metals. Acta Mater. **186**, 40–49 (2020)
33. Li, D., et al.: MOF-derived Cu_2O/Cu nanospheres anchored in nitrogen-doped hollow porous carbon framework for increasing the selectivity and activity of electrochemical CO_2-to-formate conversion. ACS Appl. Mater. Interfaces **12**(6), 7030–7037 (2020)
34. Ma, G., et al.: A hydrophobic Cu/Cu_2O sheet catalyst for selective electroreduction of CO to ethanol. Nat. Commun. **14**(1), 501 (2023)

35. Lv, X., et al.: Electron-deficient Cu sites on Cu_3Ag_1 catalyst promoting CO_2 electroreduction to alcohols. Adv. Energy Mater. **10**(37), 2001987 (2020)
36. Yang, C., et al.: Overcoming immiscibility toward bimetallic catalyst library. Sci. Adv. **6**(17), eaaz6844 (2020)
37. Dong, W.J., et al.: Tailoring electronic structure of bifunctional Cu/Ag layered electrocatalysts for selective CO_2 reduction to CO and CH_4. Nano Energy **78**, 105168 (2020)
38. Wang, M., et al.: Electrochemical reduction of CO_2 on copper-based electrocatalyst supported on MWCNTs with different functional groups. Energy Fuels (2022)
39. Platzman, I., et al.: Oxidation of polycrystalline copper thin films at ambient conditions. J. Phys. Chem. C **112**(4), 1101–1108 (2008)
40. Xue, L., et al.: Unveiling the reaction pathway on Cu/CeO_2 catalyst for electrocatalytic CO_2 reduction to CH_4. Appl. Catal. B Environ. **304**, 120951 (2022)
41. Trinh, Q.T., et al.: Synergistic application of XPS and DFT to investigate metal oxide surface catalysis. J. Phys. Chem. C **122**(39), 22397–22406 (2018)
42. Chang, C.-J., et al.: Quantitatively unraveling the redox shuttle of spontaneous oxidation/electroreduction of CuO_x on silver nanowires using in situ X-ray absorption spectroscopy. ACS Cent. Sci. **5**(12), 1998–2009 (2019)
43. Mu, S., et al.: Hydroxyl radicals dominate reoxidation of oxide-derived Cu in electrochemical CO_2 reduction. Nat. Commun. **13**(1), 3694 (2022)
44. Lian, Z., et al.: Stability and lifetime of diffusion-trapped oxygen in oxide-derived copper CO_2 reduction electrocatalysts. Nat. Catal. **7**(4), 401–411 (2024)
45. Ren, D., et al.: Tuning the selectivity of carbon dioxide electroreduction toward ethanol on oxide-derived Cu x Zn catalysts. ACS Catal. **6**(12), 8239–8247 (2016)
46. Van Muylder, J.: Electrochemical Materials Science, Chap. 1. Springer US, Boston, MA (1981)
47. Zhang, T., et al.: Tailoring* H intermediate coverage on the $CuAl_2O_4/CuO$ catalyst for enhanced electrocatalytic CO_2 reduction to ethanol. Angew. Chem. **135**(29), e202302096 (2023)
48. Zhang, Z.-Y., et al.: Cu-Zn-based alloy/oxide interfaces for enhanced electroreduction of CO_2 to C_{2+} products. J. Energy Chem. **83**, 90–97 (2023)
49. Ma, Z., et al.: Atomically dispersed Cu catalysts on sulfide-derived defective Ag nanowires for electrochemical CO_2 reduction. ACS Nano **17**(3), 2387–2398 (2023)
50. Wang, Z., et al.: Localized alkaline environment via in situ electrostatic confinement for enhanced CO_2-to-ethylene conversion in neutral medium. J. Am. Chem. Soc. (2023)
51. Xin, H., et al.: Effects of d-band shape on the surface reactivity of transition-metal alloys. Phys. Rev. B **89**(11), 115114 (2014)

Stabilized $Cu^{\delta+}$-OH Species on In Situ Reconstructed Cu Nanoparticles for CO_2-to-C_2H_4 Conversion in Neutral Media

5

5.1 Introduction

Electrochemical CO_2 reduction (ECR) into value-added chemicals and fuels using renewable energy sources holds great promise for mitigating CO_2 emissions [1, 2], but it remains at the exploratory stage. The past few decades have seen significant progress in converting CO_2 to targeted products at a lab scale [3]. However, the transitioning from lab-scale studies to highly efficient, industrial-scale production of ECR remains a formidable obstacle. The reactions in an assembled catalyst within an ECR system (e.g. flow cell or MEA) greatly deviate from the traditional ones obtained in an H-cell, introducing new complexities and dynamics that significantly affect reaction kinetics, product selectivity, and overall performance [4, 5]. Therefore, a comprehensive understanding of the mechanisms in CO_2 electrolyzer is urgently needed for the development of industrial ECR technologies.

Reactor engineering is crucial for the successful implementation of efficient and scalable ECR. Compared to flow cells, the MEA configuration leverages the gas diffusion electrode (GDE) of a flow cell, minimizing the electrolyte usage by direct contact of the catalyst layer with the ion exchange membrane. As a result, it mitigates cell resistance, boosts energy efficiency, and improves cell stability, making it more feasible for practical applications [6]. However, applying the highly efficient electrocatalysts developed in lab-scale tests in an MEA reactor often leads to significant variations in ECR performance when employing different electrolytes [5]. Specifically, although a highly alkaline environment, such as 1 M KOH or higher concentrations, increases the pH value on catalyst surfaces, suppressing the hydrogen evolution reaction (HER) and accelerating the formation of multi-carbon products [7], using strong alkaline electrolytes leads to the low CO_2 utilization efficiency and irreversible acidification of the electrolytes, posing challenges in achieving optimal performance [6, 8]. In an acidic environment, such as 0.05 M H_2SO_4

L. Wang et al., *Modulation Strategies of Cu-based Electrocatalysts for Enhancing Electrocatalytic CO_2 Conversion*, Synthesis Lectures on Green Energy and Technology,
https://doi.org/10.1007/978-3-031-98056-5_5

and 2.5 M KCl electrolytes, typical Faradaic efficiency (FE) of the HER side reaction is 20% or higher, which is energy intensive and suppresses the ECR. Additionally, the reactor corrosion in an acidic environment reduces the system's stability [9]. The drawbacks of low CO_2 utilization in alkaline electrolytes and strong HER in acid electrolytes could be significantly alleviated by using neutral electrolytes. However, the product selectivity of ECR in neutral electrolyte is lower than that in alkaline electrolyte [10, 11]. Furthermore, achieving high selectivity for ECR to C_{2+} products (e.g. ethylene and ethanol) in an MEA reactor at industrially relevant current densities represents a significant scientific and technological challenge, although state-of-the-art electrocatalysts have demonstrated remarkable progress in C_1 products (e.g., CO or HCOOH) [12, 13].

Another major obstacle to the practical application of ECR is the lack of a deep understanding of the fundamentals and mechanisms behind ECR to C_{2+} multicarbon products in an MEA reactor. Previous studies predominantly focused on catalyst design [14–18], with little attention on the real active sites in Cu-based catalysts. Thus, the real Cu active sites are still debatable due to controversial conclusions in the literature [19–22]. For example, Cu^+, mixed oxidation states ($Cu^{\delta+}$) of Cu (e.g., Cu^{2+}, Cu^+ and Cu^0) or their interface have been identified as the active sites of oxide-derived Cu for promoting CO–CO coupling under ECR [19, 20]. Conversely, another study suggests that metallic Cu is the active site of bulk Cu catalysts, attributing this to the instability of subsurface oxides under negative cathodic potentials [21]. Additionally, the active site of Cu nanoparticles (NPs) is reported to be metallic Cu nanograin, as investigated through the structural dynamics during the life cycle of Cu NPs using liquid cell scanning transmission electron microscopy [22]. However, the electron beam during in situ observations may potentially reduce the catalyst and alter the actual active sites. Furthermore, these studies primarily focus on ECR reactions in H-cells or flow cells. As a result, it remains unclear whether the proposed reaction mechanisms are applicable to MEA systems owing to the distinct differences in reactor configuration and electrode–electrolyte interface. The structural transformation and actual active sites of Cu NPs during ECR to C_{2+} products are still unrevealed in an MEA reactor at industrial-level current density.

To enhance industrial-scale CO_2-to-C_2H_4 conversion in an MEA reactor, we introduced activated carbon black (ACB) with different functional groups to manipulate the interfacial microenvironment of Cu NPs in the MEA, enhancing CO coverage to inhibit the HER and promote the desired C_{2+} reaction pathway. Using in situ multimodal characterizations, we identified the stabilized $Cu^{\delta+}$-OH species as the active sites under negative potential during ECR. The in situ-generated, strongly oxidative OH radicals stabilized the oxidized $Cu^{\delta+}$ species, leading to an irreversible and asynchronous change in morphology and valence, yielding high-curvature nanowhiskers during ECR in an MEA reactor. The well-stabilized $Cu^{\delta+}$-OH species and high-curvature nanowhiskers synergistically enhance C_2H_4 production during ECR. By understanding this mechanism, we achieved a competitive FE of 55.6% $\pm$2.8 for C_2H_4 at a current density of 316 mA cm^{-2} using a mixture of fluorinated carbon black and Cu NPs (F–Cu). The mixture of OH-functionalized carbon black and Cu NPs (OH–Cu) attained an FE of 42.1% ± 2.7 for C_2H_4 at a current density of 356.6 mA cm^{-2}. Both FEs and current densities observed with the inclusion of ACB surpass those achieved with Cu NPs only (i.e., 32.9% ± 1.8 C_2H_4 FE at a current density of 100 mA cm^{-2}). Insights from this reaction mechanism bridge the knowledge gap between lab-scale studies and industrial-scale implementation, contributing to the development of sustainable and carbon–neutral industry.

5.2 Experimental and Computational Methods

5.2.1 Materials

Cu NPs (60–80 nm in diameter, ≥ 99.5%), iridium (III) chloride hydrate ($IrCl_3 \cdot xH_2O$, 99.9%) and potassium bicarbonate ($KHCO_3$, 99.7%) were purchased from Sigma-Aldrich, Canada. Potassium hydroxide (KOH, ≥ 85%), Sodium hydroxide (NaOH, ≥ 97%), ammonium fluoride (NH_4F, ≥ 98%), oxalic acid (98%), hydrochloric acid (HCl, 37%), isopropanol (≥ 99%), and ultrapure water (18 MΩ·cm) were obtained from the University of Waterloo Chemical Store. Carbon black-Vulcan XC 72, titanium fiber felt, and Sigracet 39BB were purchased from fuel cell store, USA. Sustainion® anion exchange membrane (Sustainion® X37-50 Grade RT membrane) with 50 microns was sourced from Dioxide Materials, USA. All materials except Sustainion® X37-50 Grade RT Membrane, were used as received without further purification. Prior to use, the Sustainion® X37-50 Grade RT Membrane was cut into 4 × 4 cm pieces and activated in a 1 M KOH solution for 12 h.

5.2.2 Synthesis of Cathodic OH–Cu Electrocatalysts

For typical activation of carbon black [23], 0.2 g of carbon black was treated by 100 mL of 2 M NaOH solution with continuous stirring for 24 h at room temperature. The resulting suspension was centrifuged and washed three times with water until reaching constant pH. The synthesized sample was designated as OH-CB. The mixture of OH-CB and Cu NPs (mass ratio = 1:1) was named as OH–Cu.

5.2.3 Synthesis of Cathodic F–Cu Electrocatalysts

For the fluorination of carbon black, 0.1 g of the prepared OH-CB samples was dispersed in 70 mL isopropanol under ultrasonication, followed by addition of 0.5 g of NH_4F into the mixture. After continuous ultrasonication for 1 h, the solution was transferred into an autoclave and kept at 120 °C for 20 h. The precipitate was centrifuged, washed three times with water, and dried in a vacuum at 60 °C for 12 h. The sample was designated as F-CB. The mixture of F-CB and Cu NPs (mass ratio = 1:1) was named as F–Cu.

5.2.4 Synthesis of Anode IrO_x/Ti Electrocatalysts

IrO_x/Ti fiber felt was synthesized by a dip coating and thermal decomposition method [24]. Ti fiber felt was etched with boiling 0.5 M oxalic acid for 60 min, and 30 mg of $IrCl_3 \bullet xH_2O$ was dissolved in a 10 mL of 10% volume concentrated HCl-isopropanol solution. The Ti fiber felt obtained was then dip-coated in the above solutions, dried at 100 °C for 10 min, and calcined at 500 °C for 10 min in air. This procedure was repeated multiple times to achieve a loading of 0.5 mg.

5.2.5 Ex Situ Material Characterizations

Scanning electron microscopy (SEM, Hitachi S4800, Japan) images were acquired at an accelerating voltage of 10 kV. Glancing-incidence X-ray diffraction (GIXRD) spectra were obtained using a MRD diffractometer (PANalytical X'Pert Pro, Netherlands) with Cu Kα radiation (1.54 Å) at an incidence angle of 0.3°. XPS (Thermo-VG Scientific ESCALab 250, USA) measurements were conducted using microprobes with a monochromatic Al Kα X-ray source (1486.6 eV). High-angle annular dark-field scanning TEM (HAADF-STEM) and EDS images (FEI Titan 80-300 HB TEM/STEM, USA) were captured with an aberration corrector at an accelerating voltage of 60 kV. High resolution TEM images (Zeiss Libra 200MC, German) were collected at an acceleration voltage of 200 kV. Proton nuclear magnetic resonance (H-NMR) was run on Bruker Avance III 500 MHz, USA.

5.2.6 Electrochemical Measurements

All electrochemical measurements were connected to a potentiostat (Gamry Reference 3000 electrochemical workstation, USA) at room temperature without iR compensation. All electrochemical data were automatically saved and processed using Gamry analysis software. A home-made 5 cm^2 MEA functioned as a dual-electrode zero-gap cell, with an anion exchange membrane (Sustainion® X37-50 Grade RT membrane) sandwiched between the anode and cathode. The electrolyte circulated through the porous anode without the use of reference electrode. All applied potentials were not iR corrected. Prior to conducting the MEA test, 500 mL of 0.1 M $KHCO_3$ electrolyte ($pH = 8.3 \pm 0.2$) was prepared and stored at room temperature. For electrode fabrication, 5 mg of catalyst was mixed with 1 mL isopropanol and 20 μl Nafion solution (5 wt%, Sigma-Aldrich), followed by sonication of 99 min to form a homogenous solution. The obtained catalyst ink was air-brushed onto gas diffusion electrodes (GDEs, Sigracet 39 BB) (5 cm^2) and dried in the air. The loading of catalyst was 1mg/cm^2, and the area in contact with the membrane was 5 cm^2. IrOx/Ti fiber belt was used as the anode. The flow rate of CO_2 was controlled using a mass flow controller (0–100 sccm, Alicat Scientific, USA). The cathode side was supplied with a humidified CO_2 stream at a flow rate of 30 sccm, while the anode was circulated with 50 mL of 0.1 M $KHCO_3$ electrolyte at a flow rate of 18 mL min^{-1}. Gas products were quantified using an Agilent 6890 gas chromatograph (GC) equipped with a Carbonxen® 1000 column and a CarbonPLOT column for flame ionization detector and thermal conductivity detector, respectively. The liquid products were analyzed using a Bruker 500 MHz Nuclear Magnetic Resonance (NMR) with an internal reference of D_2O and DMSO. The FEs for gas and liquid products were calculated as follows: $FE = eF \times n/Q = eF \times n/(I \times t)$, , where e is number of transferred electrons, F is the Faraday constant, Q is the charge, I is the current, t is the testing time and n is the total amount of product (in moles).

5.2.7 In Situ XAS Measurements

The measurements were conducted at 06ID-1 Hard x-ray MicroAnalysis (HXMA) beamline of the Canadian Light Source (CLS), operated at 2.9 GeV with a constant current of 250 mA. Measurements at the copper K-edge were performed in fluorescence mode using a Lytle detector. The customized in situ electrochemical cell consists of two chambers separated by an anion exchange membrane (Sustainion® X37-50 Grade RT membrane). The cathode chamber is equipped with a thin-layer, adjustable electrolyte structure. F–Cu GDEs with a loading of 1 mg/cm^2, a platinum foil, and an Ag/AgCl electrode were used as the working electrode, counter electrode, and reference electrode, respectively. Both the cathode and anode chambers were supplied with 50 mL of CO_2-saturated 0.1 M $KHCO_3$ electrolyte (pH = 6.8 ± 0.3) at a flow rate of 18 mL/min. In situ XAS measurements was recorded at applied cathodic potential ranging from − 1 to − 3.5 V. All applied potentials were not iR corrected. The XAS data were fitting using the software package Athena.

5.2.8 In Situ Time-Dependent EPR Measurements

Time-dependent continuous-wave (CW) EPR measurements at X-band frequencies (9.4 GHz) were performed using a Bruker EMX Micro EPR spectrometer. DMPO was used as a spin trap. The electrochemical cell used for in situ EPR testing is the same as that used for the in situ XAS setup. All electrochemistry procedures are identical, except that the electrochemical test was performed in CO_2-saturated 0.1 M $KHCO_3$ electrolyte (pH

= 6.8 ± 0.3) containing 10 mM of DMPO. During ECR, 10 μL of samples were taken from the electrochemical cell and transferred into 1 mm ID, 2 mm OD EPR tube (712-SQ-100M Wilmad LabGlass). For EPR measurements, these tubes were placed into 3 mm ID 4 mm OD EPR tubes (707-SQ-250M Wilmad LabGlass). For all the measurements, the microwave power was set to 1.262 mW (22 dB microwave power attenuation) and modulation amplitude was 2 G at a modulation frequency of 100 kHz. EPR spectra were simulated using EasySpin software [25].

5.2.9 In Situ ATR FT-IR Measurements

Measurements were conducted at 02B1-1 Far-IR beamline of the Canadian Light Source (CLS). In situ ATR FT-IR spectroscopy were performed using a Bruker IFS 125 HR spectrometer with a custom-built Veemax III variable angle reflectance system (Pike Technologies, USA), and all spectra were presented as the absorbance [$-\log(R/R_0)$]. The catalyst ink was pipetted onto an ITO-coated SI Microgrooved Wafer Element with a 55° Face angle and dried in air before testing. A platinum wire and an Ag/AgCl electrode were used as counter and reference electrodes, respectively, in a customized single-chamber electrochemical cell. The 10 mL of 0.1 M $KHCO_3$ electrolyte (pH = 6.8 ± 0.3) was continuously purged with CO_2 during the measurement. The applied cathodic potentials ranged from − 0.5 to − 3.5 V, and infrared spectra were collected during ECR. All applied potentials were not iR corrected.

5.2.10 In Situ Raman Spectroscopy

Raman spectra were recorded using a Renishaw inVia Reflex system equipped with a 100 × objective and a 633 nm excitation (Renishaw HeNe laser, 17 mW). The filter was set at 5% of maximum intensity. Measurements were performed in a customized dual-chamber gas diffusion Raman electrochemical cell with a quartz window. The cathode and anode chambers were separated by an anion exchange membrane (Sustainion® X37-50 Grade RT membrane). The as-prepared catalysts on GDEs, Ag/AgCl, and platinum wire served as the working, reference, and counter electrodes, respectively. The 50 mL of 0.1 M $KHCO_3$ electrolyte (pH = 6.8 ± 0.3) was constantly purged with CO_2 gas flow rate of 30 sccm. The raw data were processed using Renishaw WiRE 5.3 software package, including baseline subtraction, normalization, and curve fitting. All applied potentials were not iR corrected.

5.2.11 The Resistance Measurements of Electrochemical Cell

Potentiostatic electrochemical impedance spectroscopy (EIS) was conducted using a potential range of − 5 to 5 V, 100 kHz frequency, and a sinus amplitude of 10 mV. The calculated cell resistances for different setups were as follows: $0.25 \pm 0.01\ \Omega$ for the MEA reactor, $212 \pm 5\ \Omega$ for the electrochemical cell used for in situ XAS and in situ EPR, $181 \pm 3\ \Omega$ for the in situ Raman electrochemical cell, and $156 \pm 2\ \Omega$ for in situ ATR FT-IR electrochemical cell. It should be noted that in this paper, all reported potentials were not iR corrected.

5.2.12 The Reference Electrode Calibration

For the typical calibration of an Ag/AgCl reference electrode [26], a three-electrode system was employed, with a Pt foil serving as both the working and the counter electrode. The 0.1 M Na_2SO_4 electrolyte was saturated with high-purity H_2 for at least 30 min before electrode calibration. Cyclic voltammetry (CV) was performed at a scan rate of 1 mV/s, and the average of two interconversion point values was taken as the thermodynamic potential.

5.3 Results

5.3.1 ECR Performance in an MEA Reactor

As-prepared samples were employed as cathode electrocatalysts in the 5 cm^2 MEA reactor (Fig. 5.1a) to evaluate the effects of ACB on ECR, including product distributions and current densities. The ECR activity of commercial Cu NPs served as the baseline for comparison with other electrocatalysts (Fig. C.1). Major products such as H_2, CO, and C_2H_4 were produced on the Cu NPs, with minimal ethanol detected. The maximum FE obtained was 32.9% ± 1.8 for C_2H_4 at ~ 100 mA cm^{-2} (Fig. C.1). However, at higher cathodic potentials, the dominant HER suppressed the C_{2+} reaction pathways, thereby limiting the partial current density of C_{2+} products. To tackle this challenge, ACB with different functional groups was introduced to regulate the interfacial microenvironment of Cu NPs in the MEA reactor. Among all the tested catalysts, F–Cu and OH–Cu exhibited high partial current densities of CO at low cell potentials (Fig. C.2a), directly influencing the subsequent C–C coupling pathway toward C_2H_4. Comparison of the partial current densities of H_2 and C_2H_4 indicated that the increased selectivity resulted from suppression of HER at high cell potentials (Fig. C.2b vs. C.2c). Therefore, at − 3.7 V, OH–Cu achieved a FE of 42.1% ± 2.7 for C_2H_4 at a current density of 356.6 mA cm^{-2} (Fig. 5.1b), while the F–Cu sample demonstrated an FE of 55.6% ± 2.8 for C_2H_4 at a current density of 316 mA cm^{-2} (Fig. 5.1c). These experimental results indicate that introducing ACB is a promising strategy to enhance the ECR performance and highlight the significant advancements in the ECR-to-C_2H_4 production, with F–Cu and OH–Cu showing excellent performance compared to most Cu-based electrocatalysts (Table 5.1).

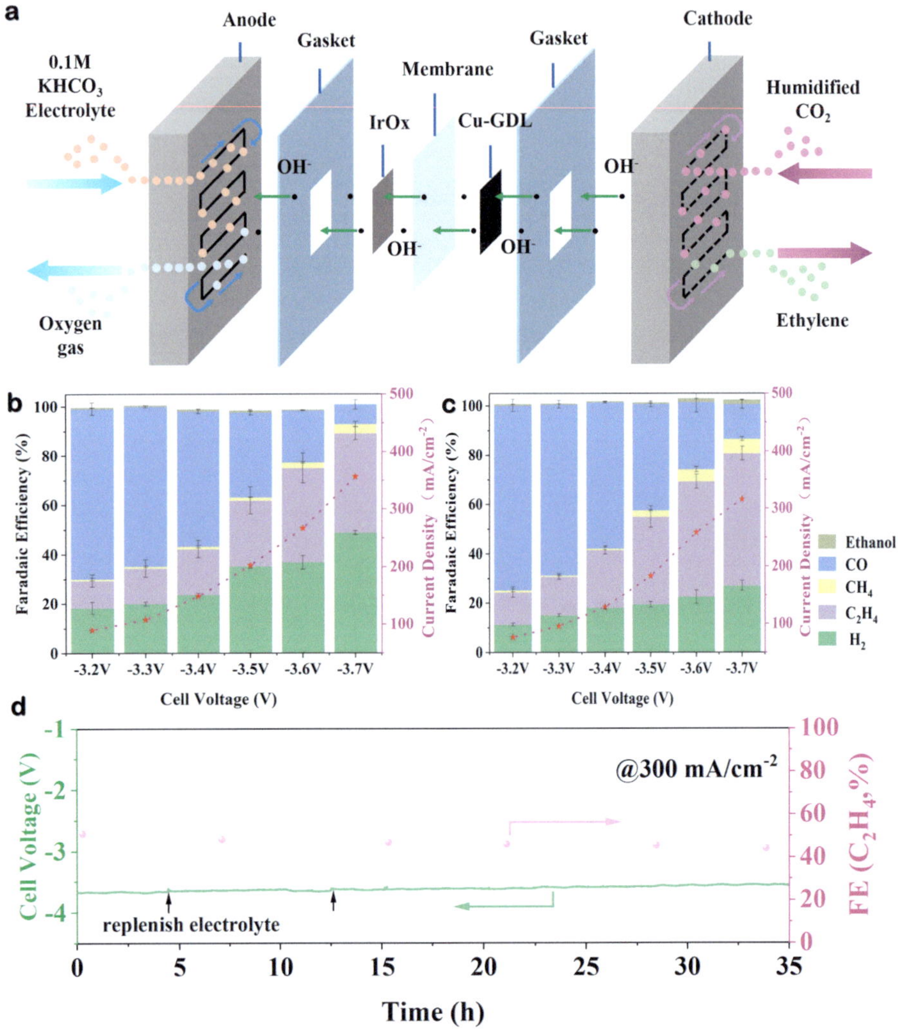

Fig. 5.1 **a** Schematic of the MEA reactor. ECR performance of **b** OH–Cu and **c** F–Cu catalysts in the 5 cm^2 MEA reactor. **d** The long-term stability test for F–Cu at current density of 300 mA cm^{-2} (1.5A). All applied potentials were not iR corrected (R = 0.25 ± 0.01 Ω). Error bars represent the standard deviations calculated from three independent measurements

Table 5.1 Comparison of optimized C_2H_4 product from various catalysts during ECR in an MEA reactor using neutral electrolyte

Electrocatalyst	Membrane	Anode	Anolyte	E (V)	j_{total} (mA cm^{-2})	FE (%)	References
F–Cu	Sustainion X37-50	IrO_x	0.1 M $KHCO_3$	– 3.7	316	55.6	This work
OH–Cu	Sustainion X37-50	IrO_x	0.1 M $KHCO_3$	– 3.7	356.6	42.1	This work
Sputtering Cu	Sustainion X37-50	IrO_2	0.1 M $KHCO_3$	– 3.9	~200	~50	[27]
Cu/N-arylpyridinium	Sustainion	IrO_x	0.1 M $KHCO_3$	– 3.65	~120	~64	[28]
Cu(100)-rich catalyst	Sustainion	IrO_x	0.15 M $KHCO_3$	– 3.7	~300	~60	[29]
Cu(100)-rich film	FAA-3-PK-75	IrO_x	0.1 M $KHCO_3$	– 2.3	120	55.8	[30]
CuNP/Cu	Sustainion	IrO_x	0.1 M $KHCO_3$	– 3.8	176	55	[31]
Carbon shell-coated Cu	Sustainion X37-50	IrO_2	0.1 M $KHCO_3$	– 3.8	~150	62.5	[32]
Thiol/Ag–Cu	FAA-3–50	IrO_x	0.1 M $KHCO_3$	– 4.5	~326	51	[33]
AP-Cu	Sustainion X37-50	IrO_x	0.1 M $KHCO_3$	– 3.8	120–160	~50	[34]
CuO-NPs@GDL	Sustainion X37-50	IrO_2	0.5 M $KHCO_3$	/	250	71.2 ± 3.3	[35]
CTPI	Sustainion X37-50	IrO_2	0.1 M $KHCO_3$	– 3.9	208	66	[36]
C/HKUST-1/Cu/PTFE	Sustainion X37-50	IrO_x	0.1 M $KHCO_3$	– 4.2	250	~50	[37]

To determine whether the catalyst's activity correlates directly with the electrochemical surface area (ECSA), we compared the double-layer capacitance (C_{dl}) of all samples. Based on the C_{dl} values, we observed that introducing ACB significantly increased the ECSA of the catalysts, thereby enhancing ECR activity (Fig. C.3). However, the ECSA of F–Cu was notably lower than that of OH–Cu. The higher activity of F–Cu indicated that ECSA cannot solely determine the enhancement of ECR activity. The fluorination promoted ECR activity by a different mechanism, such as water activation. Additionally, we varied the F-CB loading to investigate the effect of the interfacial microenvironment of Cu NPs on ECR activity (Fig. C.4). A low-loading F-CB substantially enhanced the intrinsic activity of Cu NPs, as evidenced by a higher C_2H_4 FE (42.5% $\pm$ 3.7 at a current density of 269.6 mA cm^{-2}) at − 3.6 V (Fig. C.4a) compared to Cu NPs (Fig. C.1). However, excessive F-CB loading increased the thickness of the gas diffusion electrode (GDE), thereby hindering the transport of CO_2 to the catalysts surface. Consequently, the increase in C_2H_4 FE (33% $\pm$ 4.3) at − 3.5 V was lower than that achieved with lower F-CB loading, although the current density (~200 mA cm^{-2}) (Fig. C.4b) was twice that of Cu NPs at − 3.3 V (Fig. C.1). This indicates that the modulation of the interfacial microenvironment follows a volcano-shaped curve (Fig. C.4c), necessitating the identification of an optimal loading to effectively enhance the catalyst's performance.

Long-term testing was conducted to assess the stability of the F–Cu catalyst in the 5 cm^2 MEA reactor (Fig. 5.1d). At a current density of 300 mA cm^{-2}, the catalyst operated continuously for 35 h with C_2H_4 selectivity decreasing only from 50% initially to 42%. This highlights the potential of F–Cu for industrial-relevant ECR conversion.

5.3.2 Ex Situ Characterization of Electrocatalysts

Ex situ scanning electron microscopy (SEM) images were taken to investigate the dynamic morphological and structural evolution of the catalysts. Cu NPs were sprayed onto GDEs, and their original morphology remained unchanged before the reaction (Figs. 5.2a and C.5a). After the MEA test, some nanowhiskers grew in situ on the GDEs, forming bundles and stacking on each other (Fig. C.5b, c). They had lengths of tens of micrometers and consisted of Cu, C, and O elements (Fig. C.5d, f). Similar morphological evolution and elemental distribution were observed in OH–Cu on GDEs (Fig. C.6). This indicates that carbon is involved in the structural evolution of the nanowhiskers, possibly originating from the GDEs, ACB, or $CO_3{}^{2-}$ from $KHCO_3$ electrolyte. Compared to Cu NPs, the coverage of nanowhsikers on the GDEs in the OH–Cu sample was significantly higher, and their growth was more uniform without stacking (Fig. C.5b, c vs. C.6b, c). Both samples were tested under identical experimental conditions. This suggests that the carbon originates from the ACB, triggering structural evolution. This assumption is further confirmed by the F–Cu sample (Fig. C.7). Compared to OH–Cu (Fig. C.6a–c), the longer lengths (> 50 μm) of nanowhiskers on the F–Cu sample indicate that the fluorination significantly enhances structural transformation (Figs. 5.2b, c and C.7), highlighting the major impact of functional groups on the surface of ACB on the structural evolution of nanowhiskers.

High-resolution transmission electron microscopy (HRTEM) and high-angle annular dark-field scanning TEM (HAADF-STEM) images provided deeper insights into the in situ reconstruction of Cu NPs at the nanometer scale. For example, in used F–Cu,

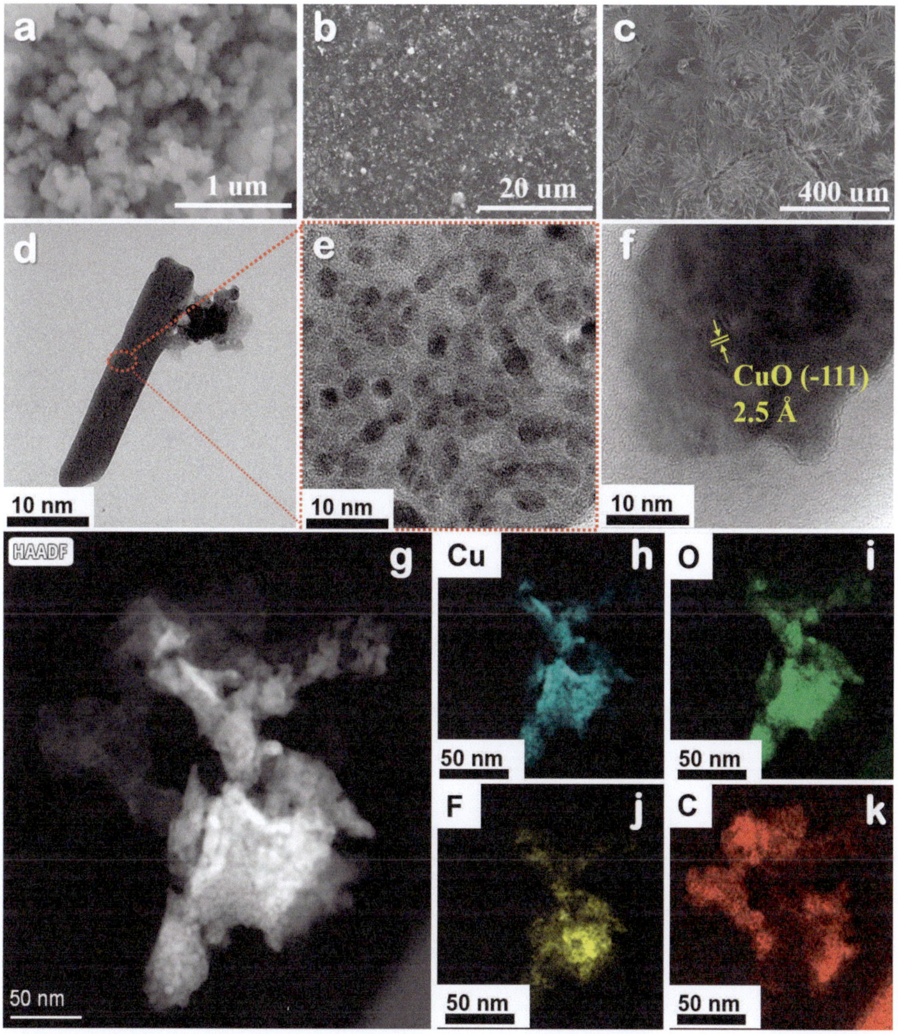

Fig. 5.2 SEM images of **a** fresh Cu NPs, **b** fresh F–Cu on GDEs, and **c** used F–Cu on GDEs. **d** TEM images and **e** corresponding high-resolution TEM images of used F–Cu catalysts; **f**, **g** HAADF-STEM images and **h–k** corresponding EDS elemental mapping images of used F–Cu catalyst

60–80 nm Cu NPs underwent potential-induced fragmentation, forming Cu nanoclusters with diameters of 3–6 nm after the ECR test in the MEA (Figs. 5.2d, e and C.8a–d). Subsequently, these nanoclusters oxidized upon exposure to air during ex situ characterization, resulting in the formation of CuO NPs (Figs. 5.2f and C.8c, d). Energy-dispersive x-ray spectroscopy (EDS) elemental mapping revealed a homogeneous distribution of Cu, O, F, and C atoms on the used F–Cu (Figs. 5.2g–k and C.9a–e). F-CB served as the substrate for the deposition of CuO NPs. Moreover, the morphological evolution of F–Cu was captured by ex situ SEM images under various cathodic potentials in the MEA. The 60–80 nm Cu NPs were held at − 3.4 V and then underwent significant particle migration at − 3.5 V (Fig. 5.3a). At this stage, the presence of nanowhiskers ~ 2.5 μm long indicated that the Cu NPs experienced initial particle reconstruction on the catalyst layer and subsequently grew into nanowhiskers with high length-to-diameter ratios (with 10 μm in length). Finally, raising the potentials promoted the growth of nanowhiskers (up to 50 μm), resulting in a uniform distribution of nanowhiskers on the surface of GDEs (Fig. 5.3a). Based on ex situ characterization analyses, we conclude that during ECR in the MEA reactor, the fragmented nanoparticles detach from the Cu matrix and rapidly evolve into well-defined nanoclusters attached to the carbon support. Subsequently, potential-induced migration of nanoclusters leads to their structural transformation into nanowhiskers [38] (Fig. 5.3a, b).

Grazing incidence X-ray diffraction (GIXRD) was performed to investigate the changes in composition and structural features of samples before and after the reaction. The characteristic peaks at 43.2°, 50.4°, and 74.1° correspond to the (111), (200) and (220) reflections of Cu, respectively (Fig. C.10a) [39]. The Cu_2O peaks at (111) and (220) in all

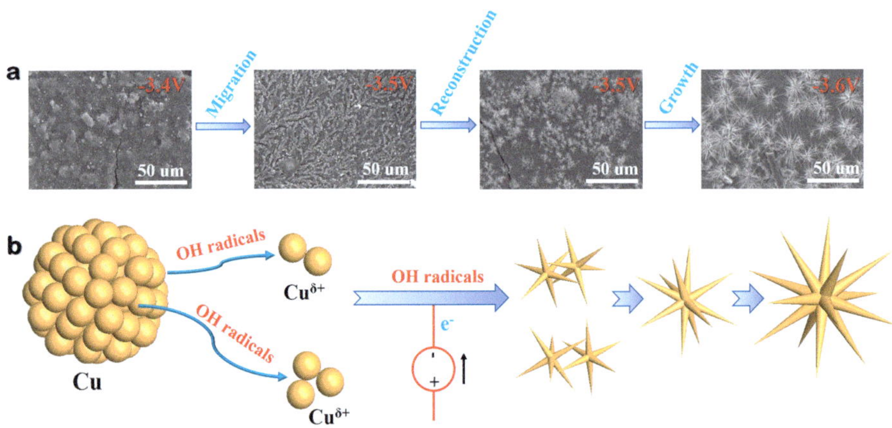

Fig. 5.3 **a** Ex situ SEM images of structural evolution of F–Cu catalysts at negative potentials of − 3.4 V, − 3.5 V and − 3.6 V in the MEA reactor. **b** A schematic structural evolution mechanism for Cu NPs during ECR in the MEA reactor. All applied potentials were not iR corrected ($R = 0.25 \pm 0.01\ \Omega$)

fresh samples indicated the oxidation of Cu [40]. GIXRD analysis revealed that before the reaction, the sample consisted of a mixture of Cu and Cu_2O (Fig. C.10a). However, after the ECR, when the sample was removed from the MEA and exposed to air for ex situ characterization, the unstable and active $Cu^{\delta+}$ species generated during ECR further oxidized to CuO. Due to the in situ growth of nanowhiskers on the GDEs, which underwent hydrophobic treatment with PTFE, and the CuO was encapsulated by the graphene-like ACB carrier formed during morphology reconstruction, it became challenging to obtain accurate phase information of the post-reaction sample (Fig. C.10b). However, conclusions drawn from STEM and TEM images in Figs. 5.2f, g and C.8c, d indicated that the intrinsic composition of the post-reaction sample was CuO loaded onto the ACB support.

X-ray photoelectron spectroscopy (XPS) was employed to investigate the surface compositions and chemical states of both fresh and used catalysts. The Cu 2p3/2 spectra of all fresh samples exhibited peaks at 933.4 and 935.2 eV, corresponding to Cu^+ or Cu^0 and Cu^{2+}, respectively (Fig. C.11a) [23]. The presence of Cu^{2+} species in all fresh samples was attributed to surface oxidation in air. The ACB significantly enhanced the intensity of the C=O bond at 532.4 eV, making it the dominant species in the O 1*s* spectra of F–Cu and OH–Cu, contrasting with the predominant lattice oxygen at 530.0 eV and the adsorbed O_2 in η_1(O) configuration (O_2^*) at 531.4 eV observed in Cu NPs (Fig. C.11b, c) [41]. After ECR, the Cu^+ or Cu^0 species were absent, and the Cu 2*p* peaks shifted to higher binding energy at ~ 935.3–936.1 eV (Fig. 5.4a). Theoretically, Cu species follow the binding energy trend: Cu^+ ions < Cu^{2+} ions < Cu(II) bound to extra-framework oxygen [42]. Since no Cu-F bond was observed at 684.4 eV in the F 1*s* spectra [43] (Fig. C.12a, b), the high binding energy in the Cu 2*p* spectra indicated that Cu species were bound to extra-framework oxygen atoms (Cu-OH). In contrast, both OH–Cu and F–Cu exhibited depletion of O_2^* species (531.4 eV) and lattice oxygen (530 eV) after the MEA test (Fig. 5.4b vs. C.11b). This revealed that lattice oxygen did not convert into O_2 during structural transformation, as evidenced by the decrease in the O_2^* peak of used Cu NPs (Fig. 5.4b). Therefore, lattice oxygen converted to OH radicals after accepting electrons from Cu atoms, leading to a shift in the Cu 2*p* binding energy to higher values (Fig. 5.4a vs. C.11a). Unlike used OH–Cu, the shift of the C=O bond peak to higher energy in used F–Cu highlighted the influence of fluorination (Fig. 5.4b). Additionally, owing to the numerous surface vacancies and hydrogen bonds, the water adsorbed at surface oxygen vacancy sites ($H_2O^*@O_{vac}$ at 533.2 eV) induced lattice rearrangement, [41] resulting in structural transformation (Fig. 5.4b).

Cu K-edge X-ray absorption Near-Edge Structure (XANES) spectra and Extended X-ray Absorption Fine Structure (EXAFS) were collected to enhance understanding of the electronic structure and coordination environment of copper species in fresh and used samples (Fig. 5.4c, d). Compared to Cu foil, XANES spectra in Fig. 5.4c indicated that the Cu sites in all fresh samples were in an oxidized state ($Cu^{\delta+}$ species, $0 < \delta < 1$). For Cu NPs, copper oxidation may be attributed to exposure to air during ex situ characterization. With OH-CB and F-CB, the absorption edge of copper shifted to higher energy,

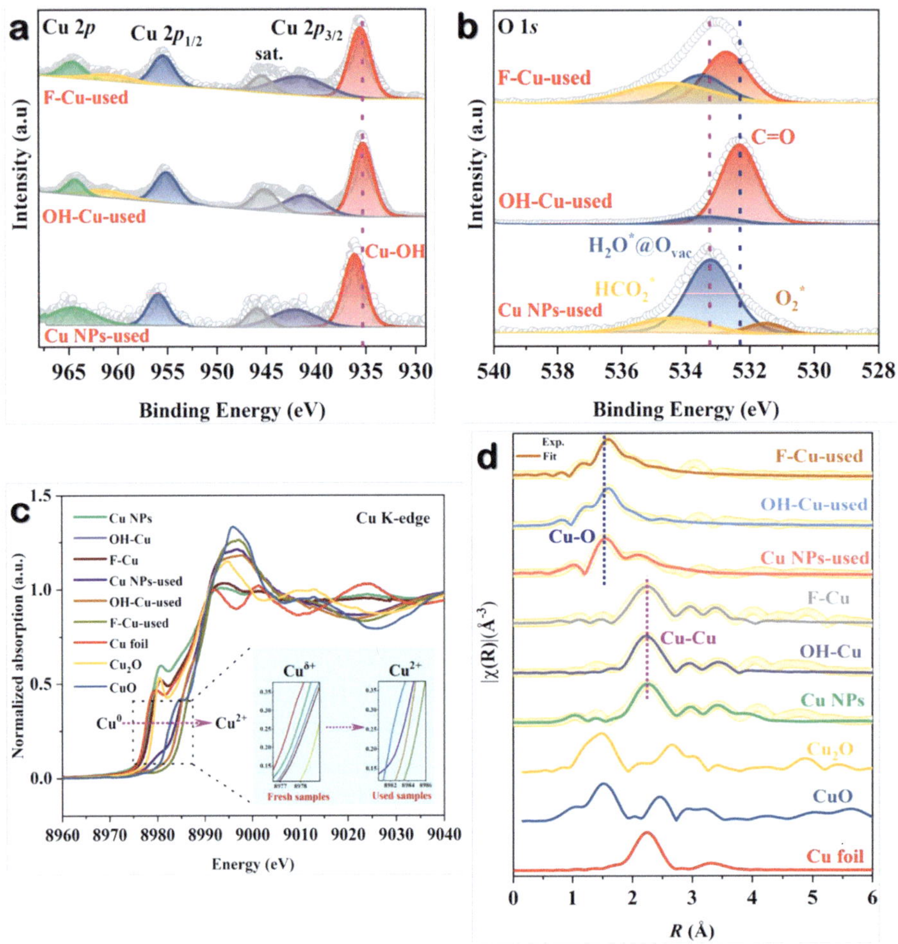

Fig. 5.4 **a** Cu 2p XPS spectra and **b** O 1s XPS spectra of used Cu NPs, OH–Cu and F–Cu catalysts; **c** Cu K-edge XANES spectra and **d** Cu K-edge EXAFS experimental and fitting spectra of fresh and used Cu NPs, OH–Cu and F–Cu samples and corresponding references (Cu foil, Cu_2O and CuO)

indicating interaction between the functional groups (OH and F) of ACB and the active copper sites. However, the oxidation state of copper in all used samples transitioned to a divalent state (Fig. 5.4c). This indicated that the ECR altered the overall structure and coordination environment of active copper sites in fresh samples, causing additional oxidation of unstable and active copper species during ECR to a divalent state upon exposed to air. Furthermore, regardless of whether samples are fresh or used, copper oxidation follows this sequence: fluorination effect > hydroxylation effect > no carbon black, highlighting the impact of functional groups on copper structure. Notably, the valence state of Cu for Cu NPs shifted to divalent after MEA testing, even in the absence of ACB. This indicated that the change in the valence state was influenced by the ECR itself, rather than by the functional groups. The functional groups (F and OH) served solely as catalysts in promoting the valence evolution. Additionally, the EXAFS fitting data of fresh Cu NPs, OH–Cu and F–Cu NPs revealed a Cu–Cu scattering path at ~ 2.52 Å (Figs. 5.4d, C.13a–c and Table C.1), and characteristic of metallic Cu. However, in fresh F–Cu, the Cu–Cu bonds transformed into Cu–O bonds after ECR testing in the MEA, reducing the coordination number from 10.3 (0.6) (Cu–Cu) before the ECR to 5.8 (0.5) (Cu–O) after ECR in the MEA (Figs. 5.4d, C.13d–f and Table C.1). Similar changes were observed in OH–Cu and Cu NPs, suggesting a uniform coordination environment among all tested samples. The oxidative species produced during ECR in the MEA destabilized active Cu sites, leading to changes in the valence state and coordination environment of Cu NPs. The functional groups from ACB (F and OH) exclusively facilitated the structural transformation of nanowhiskers.

5.3.3 In Situ Multimodal Characterizations of Electrocatalysts for ECR in Neutral Media

To enhance our understanding of ECR in an MEA reactor, in situ characterization techniques were crucial for monitoring the structural evolution of Cu species and investigating the formation of nanowhiskers. In situ Cu K-edge XANES spectra were recorded in CO_2-saturated 0.1 M $KHCO_3$ electrolyte while varying the cathodic potentials from − 1 to − 3.5 V (Figs. 5.5a, b and C.14–C.17). Switching from cyclic voltammetry (CV) to open circuit potential (OCP) caused the oxidation state of Cu sites in Cu NPs to approach the Cu(I) state ($Cu^{\delta+}$, $0 < \delta < 1$) (Fig. C.16a). This indicated that highly oxidative radicals generated in situ in the pH-neutral electrolyte facilitated the rapid reoxidation of active Cu sites. A similar phenomenon was observed in OH–Cu (Fig. C.17a) and F–Cu (Fig. 5.5a). In situ potential-dependent XANES spectra were conducted to examine the stability of $Cu^{\delta+}$ species during ECR. Despite the reduction of $Cu^{\delta+}$ species in all samples with increasing negative potentials, dynamically generated $Cu^{\delta+}$ ($0 < \delta < 1$) species persisted as non-oxide-derived Cu sites during ECR (Figs. 5.5a, C.16a and C.17a). Specifically, the $Cu^{\delta+}$ species in OH–Cu after ECR underwent a reduction and re-oxidation process in the neutral electrolyte over time (Fig. C.17a), indicating the persistence of strongly oxidative radicals in the pH-neutral electrolyte, which induced the spontaneous oxidation of Cu sites. Although Cu NPs were predominantly in their metallic state, the Cu nanoclusters formed during structural transformation were highly unstable. When exposed to air, H_2O adsorbed on surface defect sites and atmospheric oxygen can penetrate the structural network of nanowhiskers, accelerating the oxidation of $Cu^{\delta+}$ species to Cu(II) state. These findings are consistent with the observations from ex situ XANES spectra (Fig. 5.4c). Additionally, in situ Cu K-edge EXAFS spectra of Cu NPs showed mixed metallic Cu and oxidized $Cu^{\delta+}$ species at low cathodic potentials during ECR (Fig. C.16c). The strong reducing effect led to the reduction of $Cu^{\delta+}$ species, with only metallic Cu observed at high negative potentials (Fig. C.16c). As the negative potentials increased, the Cu–Cu bond length in Cu NPs gradually increased, indicating fragmentation and migration of some Cu atoms from the Cu matrix under negative bias, contributing to structural reconstruction from nanoparticles to nanowhiskers. In contrast, OH–Cu (Fig. C.17c) and F–Cu (Fig. 5.5b) exhibited stabilized $Cu^{\delta+}$ species and metallic Cu after switching to OCP. Comparing the SEM images of structural reconstruction of Cu NPs, OH–Cu, and F–Cu (Fig. C.7) revealed that the structural reconstruction of Cu NPs was primarily driven by the tensile strain effect on some metallic Cu sites resulting from the electric field and OH radicals. Hydroxylated $Cu^{\delta+}$ species play a crucial role in the large-scale and regular structural evolution of nanowhiskers in OH–Cu and F–Cu. The Cu–Cu bond length for Cu NPs was close to the initial value after ECR (Fig. C.16c), suggesting non-continuous evolution in Cu NPs. F–Cu (Fig. 5.5b) and OH–Cu (Fig. C.17c) still maintained oxidized $Cu^{\delta+}$ species after ECR, indicating continuous structural evolution in both samples, even without an electric field. Therefore, the mechanisms behind the structural evolution of Cu NPs differ from those of OH–Cu and F–Cu. For Cu NPs, the tensile strain effect induced by hydroxyl radicals detaches Cu atoms from the surface, contributing to structural transformation under negative bias. The involvement of functional groups in OH–Cu and F–Cu enhances the surface oxidative microenvironment of active Cu sites, increasing the production of oxidative radicals and stabilizing $Cu^{\delta+}$ species, thus inducing the uniform structural evolution of nanowhiskers under electric potentials.

In situ time-dependent electron paramagnetic resonance (EPR) spectra were conducted to confirm the presence of strongly oxidative radicals during ECR (Figs. 5.5c and C.18). The F–Cu catalyst underwent a 20-min ECR at − 1.5 V before electrolyte collection to mitigate the impact of residual oxidative species in the CO_2-saturated 0.1 M $KHCO_3$ electrolyte containing 10 mM 5,5-dimethyl-1-pyrroline N-oxide (DMPO). OH radicals (A_N = 1.50 mT, A_H = 1.48 mT) rather than hydrogen radicals, were produced during ECR, differing from previous studies [44]. The in situ generated strongly oxidative OH radicals can induce the oxidation of active Cu species during ECR. This indicated that a locally oxidative microenvironment on the catalyst surface facilitated the C–C coupling for C_2H_4 production, even in a neutral electrolyte. Upon switching to OCP, the OH radicals in the electrolyte did not diminish, and their persistent presence after ECR reaction stabilized active $Cu^{\delta+}$ species, further inducing structural evolution of catalysts. This finding agrees with in situ XAS spectra results (Figs. 5.5a, b and C.17).

The presence of an oxygenated Cu surface during ECR was further confirmed by the in situ Raman spectra (Fig. C.19). Ex situ SEM images tracked the morphological changes during ECR in the Raman cell, correlating with MEA reactor conditions. Similar structural evolution was confirmed (Fig. C.20). Initially, $Cu^{\delta+}$ species clustered in islands near GDEs cracks. Subsequently, structural evolution began at the center and edges of the island. Accumulated $Cu^{\delta+}$ species from the bulk phase promoted rapid nanowhiskers growth, covering the entire area. This morphological transformation highlights the relevance of our

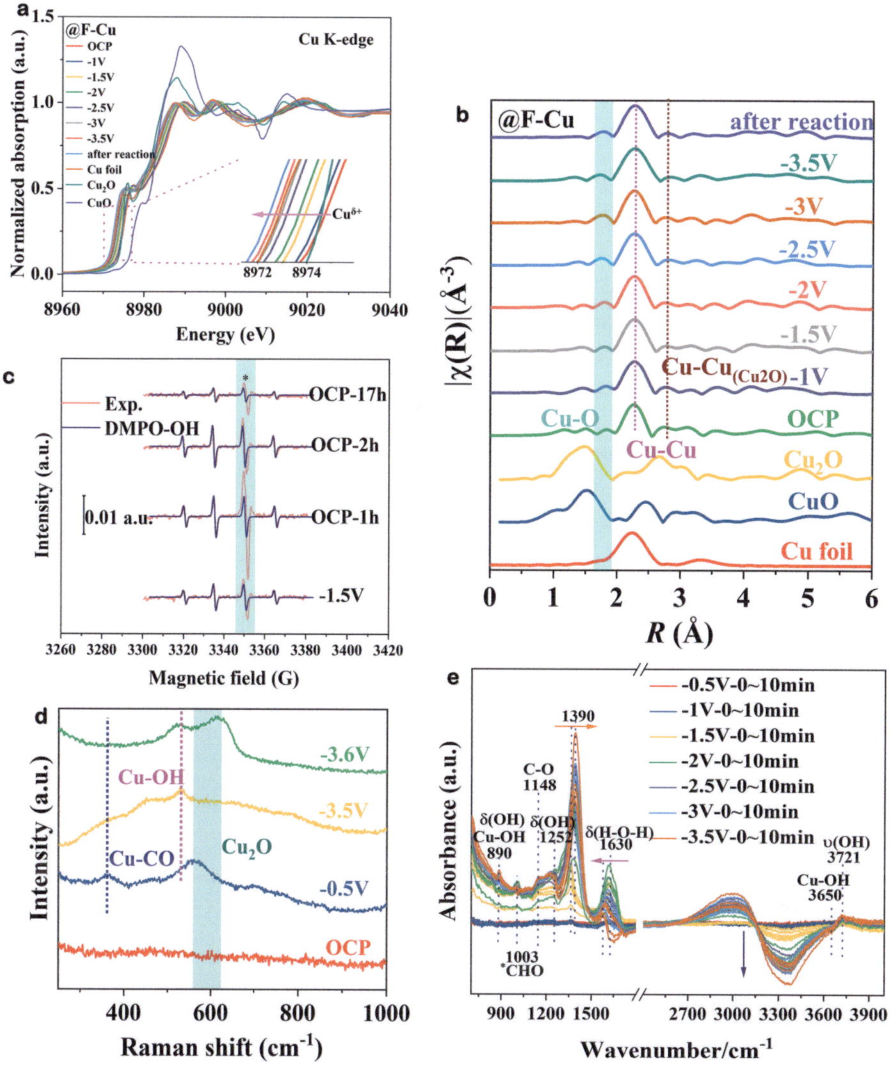

Fig. 5.5 **a** In situ Cu K-edge XANES spectra and **b** in situ Cu K-edge EXAFS spectra of F–Cu at different cathodic potentials (OCP, − 1 V, − 1.5 V, − 2 V, − 2.5 V, − 3 V, − 3.5 V, and after reaction). (R = 212 ± 5 Ω). **c** In situ time-dependent normalized EPR spectra of F–Cu in CO_2-saturated 0.1 M $KHCO_3$ electrolyte at − 1.5 V, followed by switching to OCP for different periods (R = 212 ± 5 Ω). EPR signals marked by asterisk is from the sample tube. **d** In situ Raman spectra of F–Cu in CO_2-saturated 0.1 M $KHCO_3$ electrolyte at different potentials (R = 181 ± 3 Ω). **e** In situ real-time ATR FT-IR spectra of F–Cu sample under different applied potentials (R = 156 ± 2 Ω). All applied potentials were not iR corrected

in situ Raman characterization to actual ECR in the MEA reactor. In the Raman spectra of F–Cu (Figs. 5.5d and C.21a), peaks at 565.2 cm^{-1} (Cu_2O) [44], 364 cm^{-1} (Cu–CO) [10], 1073.5 cm^{-1} (CO_3^{2-}) [44], and 1522 cm^{-1} (C=C stretching from CO_2) [45] emerged at – 0.5 V. The absence of $Cu^{\delta+}$-OH species indicated no morphology reconstruction at low potentials. However, surface $Cu^{\delta+}$-OH species at 532 cm^{-1} [10] became predominant at – 3.5 V (Fig. 5.5d), suggesting the onset of structural evolution. At – 3.6 V, a Cu_2O peak reappeared (Fig. 28d), further driving structural transformation of nanowhiskers. After ECR, these oxidized $Cu^{\delta+}$ species (Cu-OH, Cu_2O) underwent additional oxidation to form CuO (Fig. C.21b), consistent with observations from HRTEM, HADDF-STEM, and ex situ XAS spectra (Figs. 5.2f, 5.4c, d and C.8c, d). Furthermore, the resistance of $Cu^{\delta+}$ (Cu-OH, Cu_2O) species to reduction under high cathodic potentials indicates that the stabilized $Cu^{\delta+}$ species are the actual active sites during ECR, while changes in morphology enhance ECR-to-C_2H_4 conversion. Additionally, prior studies have shown that F doping can regulate the electronic structure of the active site to optimize the ECR performance [46, 47]. Therefore, we investigated the impact of F-CB by analyzing the in situ Raman spectra of Cu NPs (Fig. C.21c). We found that while Cu NPs showed Raman peaks similar to F–Cu (Cu–CO and Cu_2O) at low potentials, Cu-OH species were absent at high potentials. Comparison of the structural evolution of Cu NPs and F–Cu (Fig. C.5 vs. C.7) indicated that F-CB altered the interfacial microenvironment of Cu NPs, enhancing the structural transformation with $Cu^{\delta+}$-OH species playing a critical role. F-functional groups can activate water, enriching strongly oxidative OH radicals on the Cu surface, thereby stabilizing $Cu^{\delta+}$ species. We conclude that $Cu^{\delta+}$-OH species are not involved in reconstruction under low electric fields, resulting in a reversible process, despite the detection of $Cu^{\delta+}$ (Cu_2O and Cu–CO) species during ECR. Conversely, under high electric fields, $Cu^{\delta+}$ ($Cu^{\delta+}$-OH) species are generated; surface defects and functional groups from ACB act as kinetic trapped sites, facilitating migration of $Cu^{\delta+}$ species from bulk to support surface. In situ-generated, strongly oxidative OH radicals stabilize oxidized $Cu^{\delta+}$ species in nanowhiskers, ultimately leading to irreversible structural reconstruction under high bias.

Real-time ATR FT-IR spectra were employed to monitor intermediate adsorption processes at the electrode–electrolyte interface. Spectra were collected every two minutes during potential application and after potential cessation in a CO_2-staturated 0.1 M $KHCO_3$ solution (Figs. 5.5e, C.22 and C.23). Referring to in situ ATR FT-IR spectra of Cu NPs (Fig. C.23a), only a few intermediate peaks related to ECR were detected, likely due to multiple reaction pathways and low intermediate concentrations. In contrast, the F–Cu sample showed more prominent adsorption bands at 1003 cm^{-1} and 1148 cm^{-1}, attributed to the *CHO and the C–O bond vibration mode [48, 49], respectively. These are recognized as key intermediates in the C–C coupling during ECR to C_2H_4 (Fig. 5.5e). Furthermore, increasing cathodic potentials resulted in higher intensity of intermediate adsorption on the F–Cu surface, indicating the potential-dependent nature of ECR reaction rates. Additionally, a broad adsorption band related to H-bonded hydroxyl groups appeared in the 2500–3600 cm^{-1} range, accompanied by adsorbed water (δ(H–O–H)) at 1630 cm^{-1} (Fig. 5.5e) [50]. These bands are indicative of perturbed water molecules under negative potentials. However, extra-ligand OH exhibited stretching vibration modes of ν(OH) at 3721 cm^{-1} and δ(OH) at 1252 cm^{-1} [50, 51] on the external surfaces of F–Cu (Fig. 5.5e), primarily due to the contribution of framework modes. This indicated the presence of a metal ionic site (e.g., $Cu^{\delta+}$ species). Thus, the bands at 890 cm^{-1} and 3650 cm^{-1} were considered characteristic features of $Cu^{\delta+}$-OH species stabilized in the ν(OH) and δ(OH) modes [52, 53], respectively. Again, this underscored the crucial role of hydroxyl species as oxidants for $Cu^{\delta+}$ species (Fig. 5.5e). Although Cu NPs also exhibited infrared peaks corresponding to $Cu^{\delta+}$-OH species, there was no disturbance observed in the vibrations of water molecules at 1620 cm^{-1} (Fig. C.23a). This further indicated that the introduction of F-functional groups enhanced the activation and splitting of water molecules, facilitating Cu binding with OH species.

Intermediate adsorption processes were recorded to elucidate the origin of hydroxyl species (Figs. 5.5e and C.23b–f). Bands showing significant changes were marked with dotted lines and arrows. With increasing applied potentials, the vibrational modes of bicarbonate at 1360 cm^{-1} gradually shifted toward a broader band centered on the asymmetric C–O stretch of carbonate at 1390 cm^{-1} [54]. This was explained by the increase in OH radicals at the cathodic electrode interface during water reduction, which reduced adsorbed bicarbonate and increased carbonate under more negative potentials. This hypothesis was further confirmed by water molecule perturbations. Increasing cathodic potentials shifted δ(H–O–H) from 1630 cm^{-1} to 1575 cm^{-1} with decreased intensity, indicating that water dissociation produced more OH radicals to stabilize $Cu^{\delta+}$ species, particularly under high negative potentials. Moreover, the higher intensity of $Cu^{\delta+}$-OH and extra-ligand OH species with increasing potentials revealed changes in the surface structure. Additionally, the mechanism of depopulation of intermediates was investigated (Fig. C.23b–f). At a low cathodic potential of − 1.5 V, the bicarbonate band at 1359 cm^{-1} remained unchanged, while the δ(H–O–H) band slightly shifted. These peaks gradually disappeared after the potentials were turned off, but the in situ generated ECR intermediates disappeared intermediately. Further increasing the potential resulted in a greater shift for bicarbonate and stronger perturbations of water molecules, but they gradually returned to the initial stage when the potentials were removed. Furthermore, extra-ligand OH species were unlikely to stabilize on active $Cu^{\delta+}$ sites after ECR under high negative potentials. Instead, water molecules coordinated with $Cu^{\delta+}$ species to form aquo-complexes and maintained a specific configuration. When exposed to air, it gradually oxidized to CuO.

Based on the above analysis, OH radicals can create a locally oxidative microenvironment on the Cu surface during ECR, stabilizing $Cu^{\delta+}$ species and enhancing C_2H_4 selectivity. Although in situ Raman spectroscopy, EPR spectra, ATR FT-IR spectroscopy, and XAS spectra can directly detect $Cu^{\delta+}$-OH species, the significant difference in current densities between in situ electrochemical cell and MEA reactor precludes accurate simulation of catalyst evolution in an MEA. This is because the rapid kinetics of the ECR at high current densities introduce gas bubble that critically affect data acquisition. Furthermore, in the MEA reactor, the catalyst layer on the GDEs is closely attached to the AEM, preventing direct monitoring of surface structural changes of catalyst using Raman, infrared, and EPR spectroscopy. XAS spectroscopy could be a feasible method for monitoring changes in the oxidation state and coordination environment of copper during the MEA testing. However, to gain more structural insights, future endeavors should focus on rationally designing in situ MEA cells and conducting coordinated multimodal characterization studies to elucidate the evolution of active sites.

5.4 Conclusion

In summary, functionalized ACB was introduced to modulate the interfacial microenvironment of Cu NPs in the pH-neutral electrolyte. The enhanced performance of ECR to C_2H_4 (55.6% $\pm$ 2.8 FE at 316 mA cm^{-2}) in an MEA reactor was attributed to the locally oxidative microenvironment on the catalyst surface, regulated by hydroxyl species generated in situ from water dissociation. The well-stabilized $Cu^{\delta+}$-OH species induced the fragmentation of Cu NPs from Cu matrix and their migration on the carbon support under high negative bias, resulting in an irreversible and asynchronous change in morphology and valence, yielding high-curvature structure (nanowhiskers). Understanding the reaction mechanism in a MEA electrolyzer would better guide the redesign and assembly of catalysts on the GDEs within MEA, accelerating the commercialization of ECR.

Competing Interests The authors have no conflicts of interests to declare that are relevant to the content of this chapter.

References

1. Bushuyev, O.S., et al.: What should we make with CO_2 and how can we make it? Joule. **2**(5), 825–832 (2018)
2. Nitopi, S., et al.: Progress and perspectives of electrochemical CO_2 reduction on copper in aqueous electrolyte. Chem. Rev. **119**(12), 7610–7672 (2019)
3. Zheng, T., et al.: Large-scale and highly selective CO_2 electrocatalytic reduction on nickel single-atom catalyst. Joule. **3**(1), 265–278 (2019)
4. Popović, S., et al.: Stability and degradation mechanisms of copper-based catalysts for electrochemical CO_2 reduction. Angew. Chem. **132**(35), 14844–14854 (2020)
5. Fan, L., et al.: Strategies in catalysts and electrolyzer design for electrochemical CO_2 reduction toward C_{2+} products. Sci. Adv. **6**(8), eaay3111 (2020)
6. Chen, C., et al.: Address the "alkalinity problem" in CO_2 electrolysis with catalyst design and translation. Joule. **5**(4), 737–742 (2021)

7. Dinh, C.-T., et al.: CO_2 electroreduction to ethylene via hydroxide-mediated copper catalysis at an abrupt interface. Science **360**(6390), 783–787 (2018)
8. Zhang, X., et al.: Selective and high current CO_2 electro-reduction to multicarbon products in near-neutral KCl electrolytes. J. Am. Chem. Soc. **143**(8), 3245–3255 (2021)
9. Cao, Y., et al.: Surface hydroxide promotes CO_2 electrolysis to ethylene in acidic conditions. Nat. Commun. **14**(1), 2387 (2023)
10. Wang, Z., et al.: Localized alkaline environment via in situ electrostatic confinement for enhanced CO_2-to-ethylene conversion in neutral medium. J. Am. Chem. Soc. (2023)
11. Ma, W., et al.: Electrocatalytic reduction of CO_2 to ethylene and ethanol through hydrogen-assisted C-C coupling over fluorine-modified copper. Nat. Catal. **3**(6), 478–487 (2020)
12. Endrődi, B., et al.: Continuous-flow electroreduction of carbon dioxide. Progr. Energy Combust. Sci. **62**, 133–154 (2017)
13. Zheng, T., et al.: Copper-catalysed exclusive CO_2 to pure formic acid conversion via single-atom alloying. Nat. Nanotechnol., 1–8 (2021)
14. Xie, M.S., et al.: Amino acid modified copper electrodes for the enhanced selective electroreduction of carbon dioxide towards hydrocarbons. Energy Environ. Sci. **9**(5), 1687–1695 (2016)
15. Zhou, Y., et al.: Dopant-induced electron localization drives CO_2 reduction to C_2 hydrocarbons. Nat. Chem. **10**(9), 974–980 (2018)
16. Ledezma-Yanez, I., et al.: Structure-sensitive electroreduction of acetaldehyde to ethanol on copper and its mechanistic implications for CO and CO_2 reduction. Catal. Today **262**, 90–94 (2016)
17. Lei, Q., et al.: Structural evolution and strain generation of derived-Cu catalysts during CO_2 electroreduction. Nat. Commun. **13**(1), 4857 (2022)
18. Lv, X., et al.: Electron-deficient Cu sites on Cu_3Ag_1 catalyst promoting CO_2 electroreduction to alcohols. Adv. Energy Mater. **10**(37), 2001987 (2020)
19. Mistry, H., et al.: Highly selective plasma-activated copper catalysts for carbon dioxide reduction to ethylene. Nat. Commun. **7**(1), 1–9 (2016)
20. Liang, Z.-Q., et al.: Copper-on-nitride enhances the stable electrosynthesis of multi-carbon products from CO_2. Nat. Commun. **9**(1), 1–8 (2018)
21. Lum, Y., Ager, J.W.: Stability of residual oxides in oxide-derived copper catalysts for electrochemical CO_2 reduction investigated with 18O labeling. Angew. Chem. Int. Ed. **57**(2), 551–554 (2018)
22. Yang, Y., et al.: Operando studies reveal active Cu nanograins for CO_2 electroreduction. Nature **614**(7947), 262–269 (2023)
23. Wang, M., et al.: Electrochemical reduction of CO_2 on copper-based electrocatalyst supported on MWCNTs with different functional groups. Energy Fuels (2022)
24. Luc, W., et al.: An Ir-based anode for a practical CO_2 electrolyzer. Catal. Today **288**, 79–84 (2017)
25. Stoll, S., Schweiger, A.: EasySpin, a comprehensive software package for spectral simulation and analysis in EPR. J. Magn. Reson. **178**(1), 42–55 (2006)
26. Niu, S., et al.: How to reliably report the overpotential of an electrocatalyst. ACS Energy Lett. **5**(4), 1083–1087 (2020)
27. Gabardo, C.M., et al.: Continuous carbon dioxide electroreduction to concentrated multi-carbon products using a membrane electrode assembly. Joule. **3**(11), 2777–2791 (2019)
28. Li, F., et al.: Molecular tuning of CO_2-to-ethylene conversion. Nature **577**(7791), 509–513 (2020)
29. Wang, Y., et al.: Catalyst synthesis under CO_2 electroreduction favours faceting and promotes renewable fuels electrosynthesis. Nat. Catal. **3**(2), 98–106 (2020)

30. Zhang, G., et al.: Efficient CO_2 electroreduction on facet-selective copper films with high conversion rate. Nat. Commun. **12**(1), 5745 (2021)
31. Xu, Y., et al.: Self-cleaning CO_2 reduction systems: unsteady electrochemical forcing enables stability. ACS Energy Lett. **6**(2), 809–815 (2021)
32. Kim, J.-Y., et al.: Quasi-graphitic carbon shell-induced Cu confinement promotes electrocatalytic CO_2 reduction toward C_{2+} products. Nat. Commun. **12**(1), 3765 (2021)
33. Wu, H., et al.: Improved electrochemical conversion of CO_2 to multicarbon products by using molecular doping. Nat. Commun. **12**(1), 7210 (2021)
34. Zhang, J., et al.: Surface promotion of copper nanoparticles with alumina clusters derived from layered double hydroxide accelerates CO_2 reduction to ethylene in membrane electrode assemblies. Nano Res., 1–6 (2022)
35. Liu, W., et al.: Electrochemical CO_2 reduction to ethylene by ultrathin CuO nanoplate arrays. Nat. Commun. **13**(1), 1877 (2022)
36. Ozden, A., et al.: High-rate and efficient ethylene electrosynthesis using a catalyst/promoter/transport layer. ACS Energy Lett. **5**(9), 2811–2818 (2020)
37. Nam, D.H., et al.: High-rate and selective CO_2 electrolysis to ethylene via metal–organic-framework-augmented CO_2 availability. Adv. Mater. **34**(51), 2207088 (2022)
38. Li, F., et al.: Interplay of electrochemical and electrical effects induces structural transformations in electrocatalysts. Nat. Catal. **4**(6), 479–487 (2021)
39. Gu, Z., et al.: Oxygen vacancy tuning toward efficient electrocatalytic CO_2 reduction to C_2H_4. Small Methods. **3**(2), 1800449 (2019)
40. Li, D., et al.: MOF-derived Cu_2O/Cu nanospheres anchored in nitrogen-doped hollow porous carbon framework for increasing the selectivity and activity of electrochemical CO_2-to-formate conversion. ACS Appl. Mater. Interfaces **12**(6), 7030–7037 (2020)
41. Trinh, Q.T., et al.: Synergistic application of XPS and DFT to investigate metal oxide surface catalysis. J. Phys. Chem. C **122**(39), 22397–22406 (2018)
42. Artiglia, L., et al.: In Situ X-ray photoelectron spectroscopy detects multiple active sites involved in the selective anaerobic oxidation of methane in copper-exchanged zeolites. ACS Catal. **9**(8), 6728–6737 (2019)
43. Sun, L., et al.: Interface-structure-modulated CuF_2/CF x composites for high-performance lithium primary batteries. Energy Environ. Mater. **6**(2), e12323 (2023)
44. Mu, S., et al.: Hydroxyl radicals dominate reoxidation of oxide-derived Cu in electrochemical CO_2 reduction. Nat. Commun. **13**(1), 3694 (2022)
45. Zhao, Y., et al.: Elucidating electrochemical CO_2 reduction reaction processes on Cu (hkl) single-crystal surfaces by in situ Raman spectroscopy. Energy Environ. Sci. **15**(9), 3968–3977 (2022)
46. Ko, Y.-J., et al.: Exploring dopant effects in stannic oxide nanoparticles for CO_2 electro-reduction to formate. Nat. Commun. **13**(1), 2205 (2022)
47. Xie, J., et al.: Metal-free fluorine-doped carbon electrocatalyst for CO_2 reduction outcompeting hydrogen evolution. Angew. Chem. Int. Ed. **57**(31), 9640–9644 (2018)
48. Yi, J.D., et al.: Highly selective CO_2 electroreduction to CH_4 by in situ generated Cu_2O single-type sites on a conductive MOF: stabilizing key intermediates with hydrogen bonding. Angew. Chem. Int. Ed. **59**(52), 23641–23648 (2020)
49. Xie, H., et al.: Achieving highly selective electrochemical CO_2 reduction to C_2H_4 on Cu nanosheets. J. Energy Chem. **79**, 312–320 (2023)
50. Borfecchia, E., et al.: Revisiting the nature of Cu sites in the activated Cu-SSZ-13 catalyst for SCR reaction. Chem. Sci. **6**(1), 548–563 (2015)
51. Zhang, R., et al.: Photocatalytic oxidative dehydrogenation of ethane using CO_2 as a soft oxidant over Pd/TiO_2 catalysts to C_2H_4 and syngas. ACS Catal. **8**(10), 9280–9286 (2018)

52. Iwasaki, M., Shinjoh, H.: NO evolution reaction with NO_2 adsorption over Fe/ZSM-5: In situ FT-IR observation and relationships with Fe sites. J. Catal. **273**(1), 29–38 (2010)
53. Giordanino, F., et al.: Characterization of Cu-exchanged SSZ-13: a comparative FTIR, UV-Vis, and EPR study with Cu-ZSM-5 and Cu-β with similar Si/Al and Cu/Al ratios. Dalton Trans. **42**(35), 12741–12761 (2013)
54. Wang, L., et al.: In situ spectroscopic monitoring of CO_2 reduction at copper oxide electrode. Faraday Discuss. **197**, 517–532 (2017)

Conclusion and Outlook

6

6.1 Conclusion

This book explores modulation strategies for Cu-based catalysts to control the reduction products in CO_2 electrolysis. The experimental research ranges from lab-scale H-cells to industrial-scale MEA reactors under neutral conditions. Mechanistic insights reveal that, in neutral H-cells, among the complex structure assembles of alloying, oxidation states, and GBs, nanoalloy systems tend to favor methane production, while oxidized $Cu^{\delta+}$ species serve as real active sites that effectively promote C_{2+} product formation. To stabilize oxidized Cu species at high current densities in an MEA reactor, we use activated carbon black to modulate the catalyst's interface microenvironment. In situ generated strongly oxidative hydroxyl species can create a locally oxidative microenvironment on the catalyst surface, stabilizing $Cu^{\delta+}$ species. The proposed design strategies are universally applicable and aid in the transition from laboratory ECR research to large-scale industrial applications.

(1) In CuAg systems rich in oxidation states and grain boundary structures, the true active sites are still debatable, and their role in governing CH_4 conversion remains unclear. Herein, this study answers these questions using bimetallic catalysts by novel electro-shock strategy with controllable GBs for the oxidization of $Cu^{\delta+}$ species by modulating Ag loading. The Ag enrichment at the GBs facilitates the bonding of oxygen with the uncoordinated Cu atoms, resulting in GBO effect. The obtained CH_4 selectivity is twice that of GBs effect. The enhanced performance is attributed to the stable $Cu^{\delta+}$ species, unique electron transfer mechanism, and strong tensile Cu–O bond strain from GBO structure. In situ attenuated-total-reflection Fourier-transform-infrared-spectroscopy unveils the reaction pathway of CO_2-to-CH_4 and the sluggish reversible quenching processes of intermediates. Theoretical calculations indicate that

L. Wang et al., *Modulation Strategies of Cu-based Electrocatalysts for Enhancing Electrocatalytic CO_2 Conversion*, Synthesis Lectures on Green Energy and Technology,
https://doi.org/10.1007/978-3-031-98056-5_6

the weak $^{*}CO$ adsorption on GBO structure facilitates $^{*}CO$ hydrogenation, promoting CO_2-to-CH_4 conversion.

(2) Although intricate structural assemblies contribute to enhancing activity of ECR to C_{2+} products, blindly coupling multiple design strategies may not yield the expected results, and even inhibit the activity of intrinsic catalytic sites. Therefore, elucidating the promoting or inhibitory effects of each design strategy on the CO_2-to-C_{2+} conversion to clarify the real active sites and dynamic oxidation processes is of paramount importance. Here, we focus on commonly used GBs, oxidation states, and alloying strategies, constructing four different types of catalysts structures: original Cu GBs, oxygen-enriched GBO, Ag-enriched GBO, and Cu/Ag GBs. Multiple in situ characterizations reveal that GBs and GBO strengthen resistance of the oxidative Cu species to the electrochemical reduction. The in situ generated strongly oxidative hydroxyl radicals alter local reaction environment on the catalyst surface, inducing and stabilizing oxidative $Cu^{\delta+}$ species. Catalytic activity comparisons indicate that the oxidation state of Cu plays a decisive role in the CO_2-to-C_{2+} conversion, and nanoalloy effect tends to favor the CH_4 production in intricate GBs assemblies. Theoretical calculations suggest that weak CO adsorption on GBO structures facilitates hydrogenation, promoting C–C coupling towards C_{2+} products.

(3) Achieving large-scale ECR to multicarbon products with high selectivity using MEA electrolyzers in neutral electrolyte is promising for carbon neutrality. However, the unsatisfactory multicarbon products selectivity and unclear reaction mechanisms in an MEA have hindered its further development. Here, for upscaling CO_2 conversion technology, we report a strategy that manipulates the interfacial microenvironment of Cu nanoparticles in an MEA to suppress hydrogen evolution reaction and enhance C_2H_4 conversion. In situ multimodal characterizations consistently reveal well-stabilized $Cu^{\delta+}$-OH species as active sites during MEA testing. The OH radicals generated in situ from water create a locally oxidative microenvironment on the copper surface, stabilizing the $Cu^{\delta+}$ species and leading to an irreversible and asynchronous change in morphology and valence, yielding high-curvature nanowhiskers. Consequently, we deliver a selective C_2H_4 production with a Faradaic efficiency of $55.6\% \pm 2.8$ at 316 mA cm^{-2} in neutral media.

6.2 Outlook

Despite extensive efforts in ECR systems, several issues remain that must be addressed before commercialization. These issues primarily revolve product selectivity, stability, understanding of reaction mechanisms, and energy efficiency. The specific details are as follows:

(1) While C_1 products like CO and formate can achieve nearly 100% selectivity at high current densities, the selectivity for multi-carbon products remains insufficient for industrial applications. Although multi-carbon products could ideally serve as direct fuels, their high separation costs and the mixture of various gaseous and liquid products make them impractical for continuous supply in downstream reactions. Future efforts should focus on designing catalysts that improve the selectivity for single products, making them more practical for industrial applications.
(2) Industrial-scale ECR technology demands long-term stability. Challenges such as catalyst deactivation, the electrowetting effect of the GDL, and membrane durability hinder the advancement of ECR technology. Therefore, ensuring the long-term operation of Cu-based catalysts in reactors remains a significant challenge. Anchoring catalysts onto superhydrophobic GDLs and developing high-performance ion-exchange membranes could improve the stability of CO_2 electrolyzers.
(3) Previous mechanistic studies have generally concentrated on H-cell setups at the lab scale, which may not be applicable to flow cells and MEA reactors. The significant differences in current densities and configurations between flow cells, MEA reactors, and H-cells indicate that conventional in situ reaction cells may not accurately reflect the reaction processes in flow cells and MEAs. Therefore, developing appropriate in situ flow cell and MEA systems is essential for elucidating the true reaction mechanisms under realistic ECR conditions.
(4) The overall energy efficiency of CO_2 electrolyzers depends on the cell voltage. Optimizing the reactor's structural design to lower the cell voltage is a viable approach. Techno-economic analyses should be conducted to evaluate the practicality of CO_2 electrolyzers for practical applications. Economic cost evaluations will demonstrate the significant potential of CO_2 conversion technologies.

In summary, although significant progress has been made, addressing these key issues is crucial for the commercial viability of ECR technologies. Focusing on these areas will facilitate the transition of ECR from laboratory research to large-scale industrial applications.

Competing Interests The authors have no conflicts of interests to declare that are relevant to the content of this chapter.

Appendix A: Supporting Information for Chap. 3

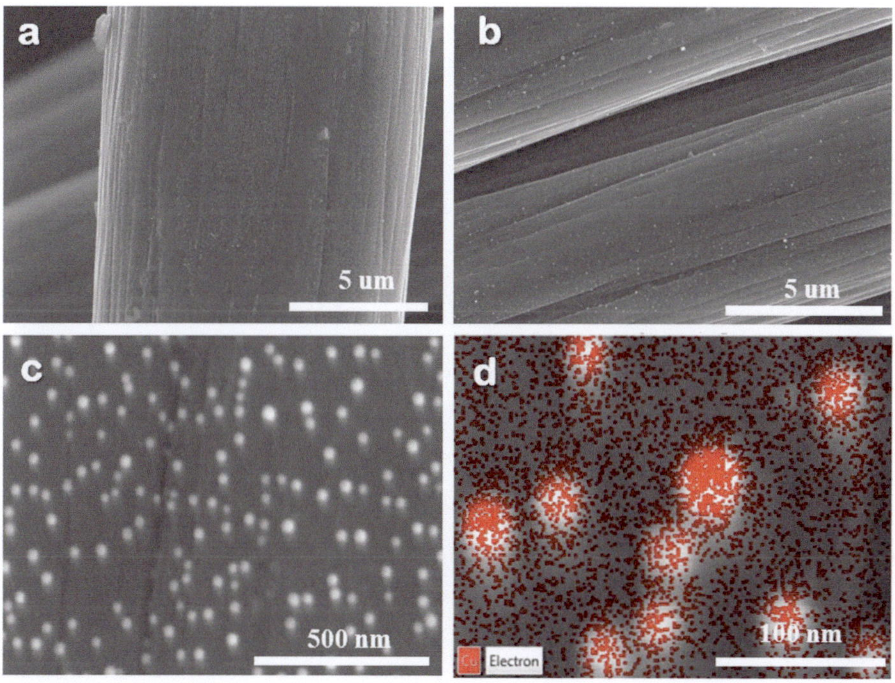

Fig. A.1 Effect of concentration on the size of ES-Nanoparticles. **a–c** SEM images and **d** corresponding EDS element mapping image of ES-Cu-0.1 mM catalyst synthesized by 0.1 mM Cu metal precursor solution

L. Wang et al., *Modulation Strategies of Cu-based Electrocatalysts for Enhancing Electrocatalytic CO_2 Conversion*, Synthesis Lectures on Green Energy and Technology,
https://doi.org/10.1007/978-3-031-98056-5

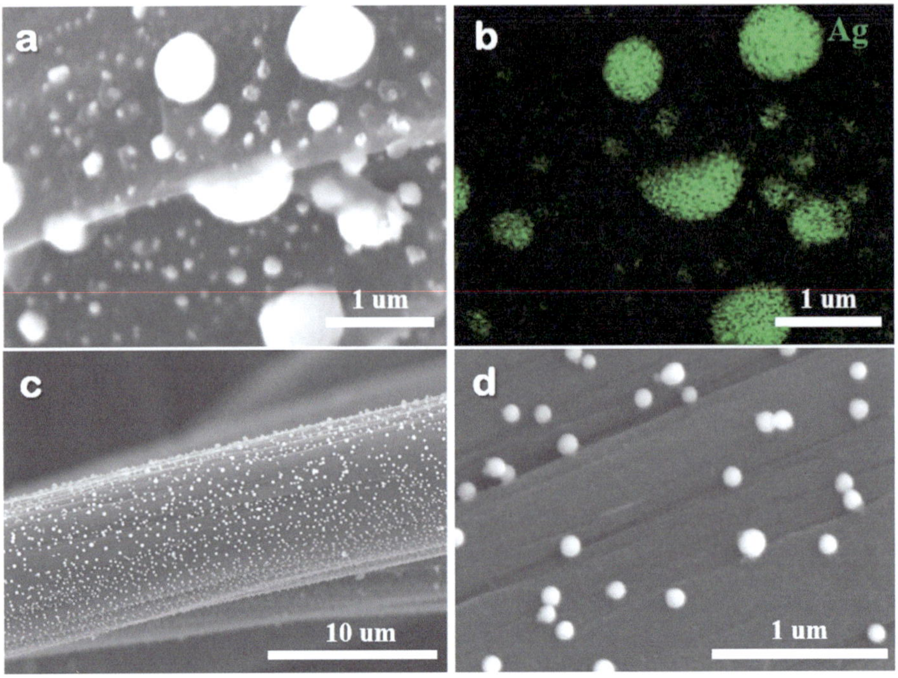

Fig. A.2 **a** SEM images and **b** corresponding EDS element mapping image of ES-Ag sample. **c**, **d** SEM images of ES-Cu sample

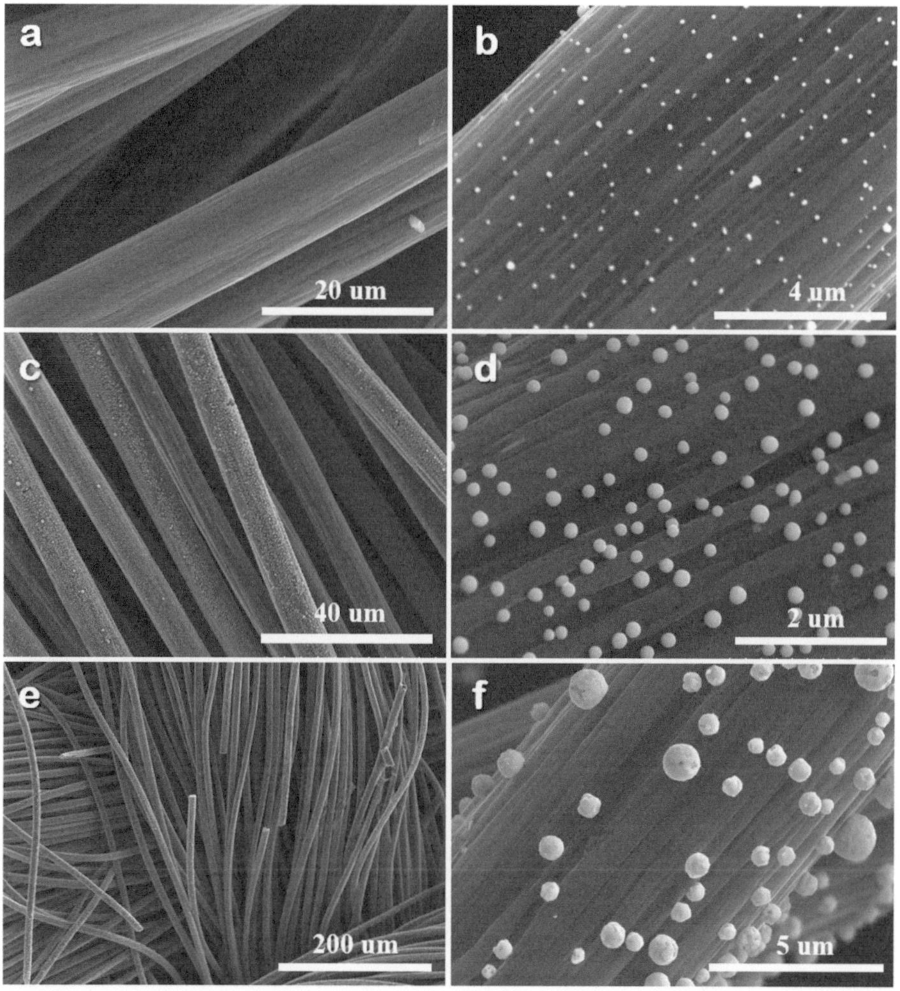

Fig. A.3 **a**, **b** SEM images of ES-50%CuAg-0.05s sample synthesized by 0.05 s electro-shock reduction. **c**, **d** SEM images of ES-50%CuAg-0.2s sample synthesized by 0.2 s electro-shock reduction. **e**, **f** SEM images of ES-50%CuAg-0.5s sample synthesized by 0.5s electro-shock reduction

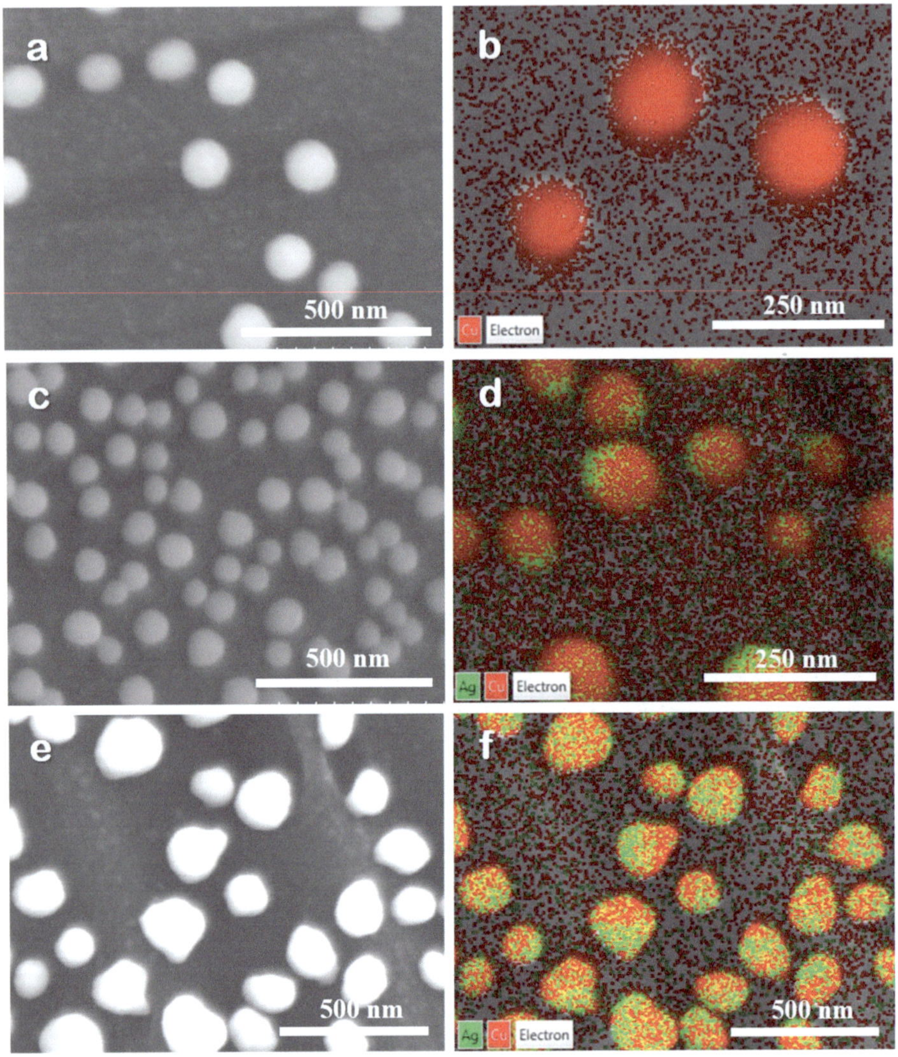

Fig. A.4 SEM images and corresponding EDS element mapping image of **a**, **b** ES-Cu-0.1s, **c**, **d** ES-1%CuAg-0.1s, and **e**, **f** ES-50%CuAg-0.1s samples

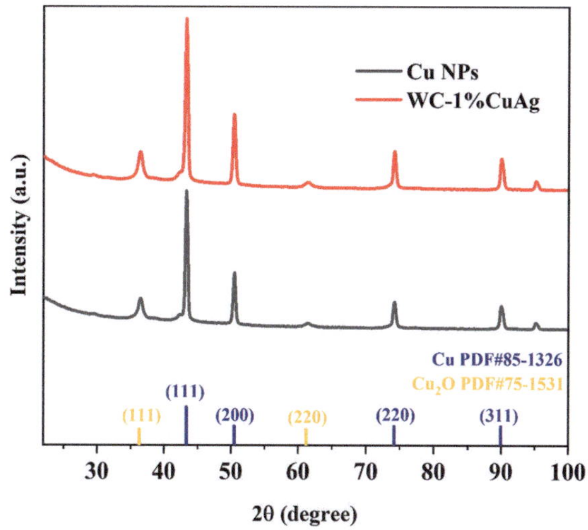

Fig. A.5 GIXRD pattern of Cu NPs and WC-1%CuAg NPs

Table A.1 The metal content measured by inductively coupled plasma mass spectrometry (ICP-MS)

Catalyst	Cu (ng/g)	Ag (ng/g)	Ag content (wt%)
ES-Cu	580,037	/	/
ES-1%CuAg	678,955	11,870	1.7
ES-50%CuAg	620,028	294,513	47.5

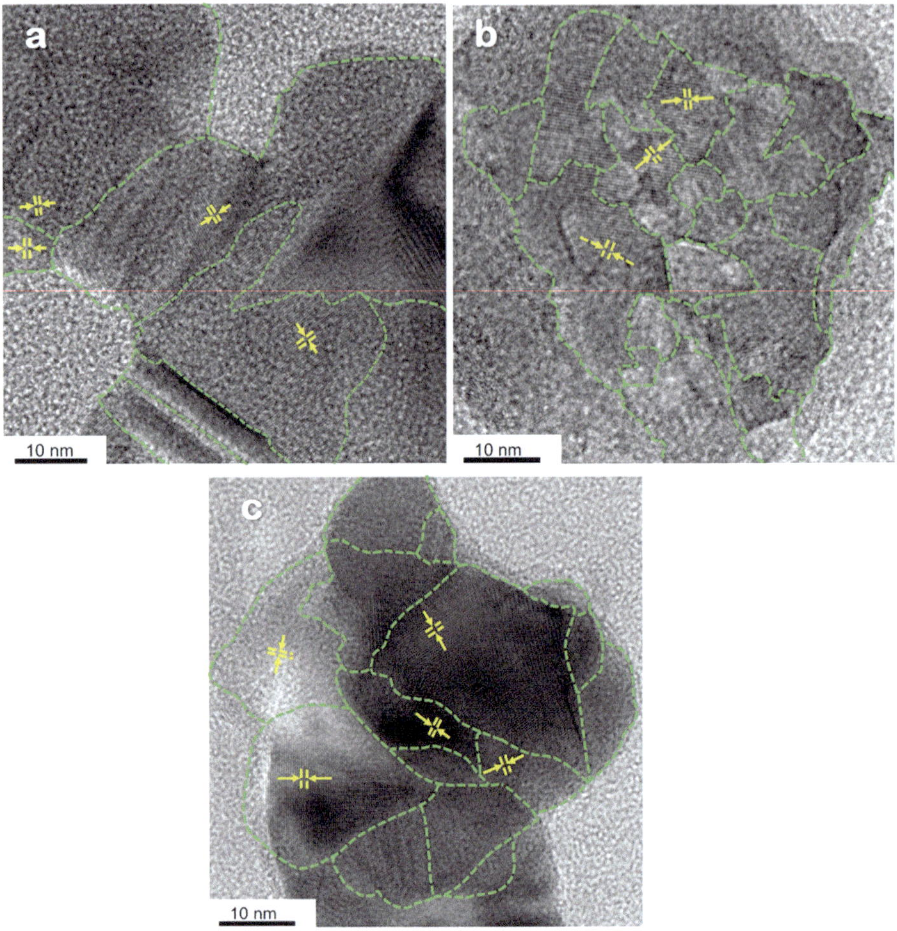

Fig. A.6 HRTEM images of grain boundaries in **a** ES-Cu, **b** ES-1%CuAg and **c** ES-50%CuAg samples

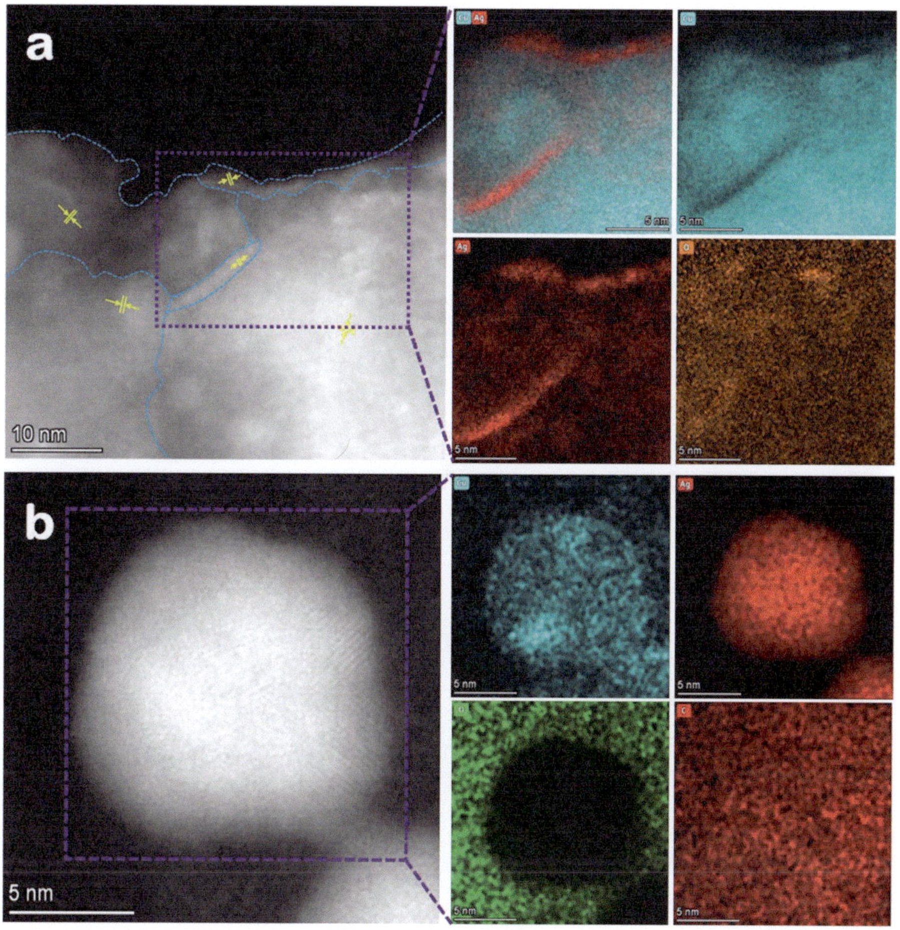

Fig. A.7 STEM-EDX element mapping image of **a** ES-1%CuAg and **b** ES-50%CuAg, respectively

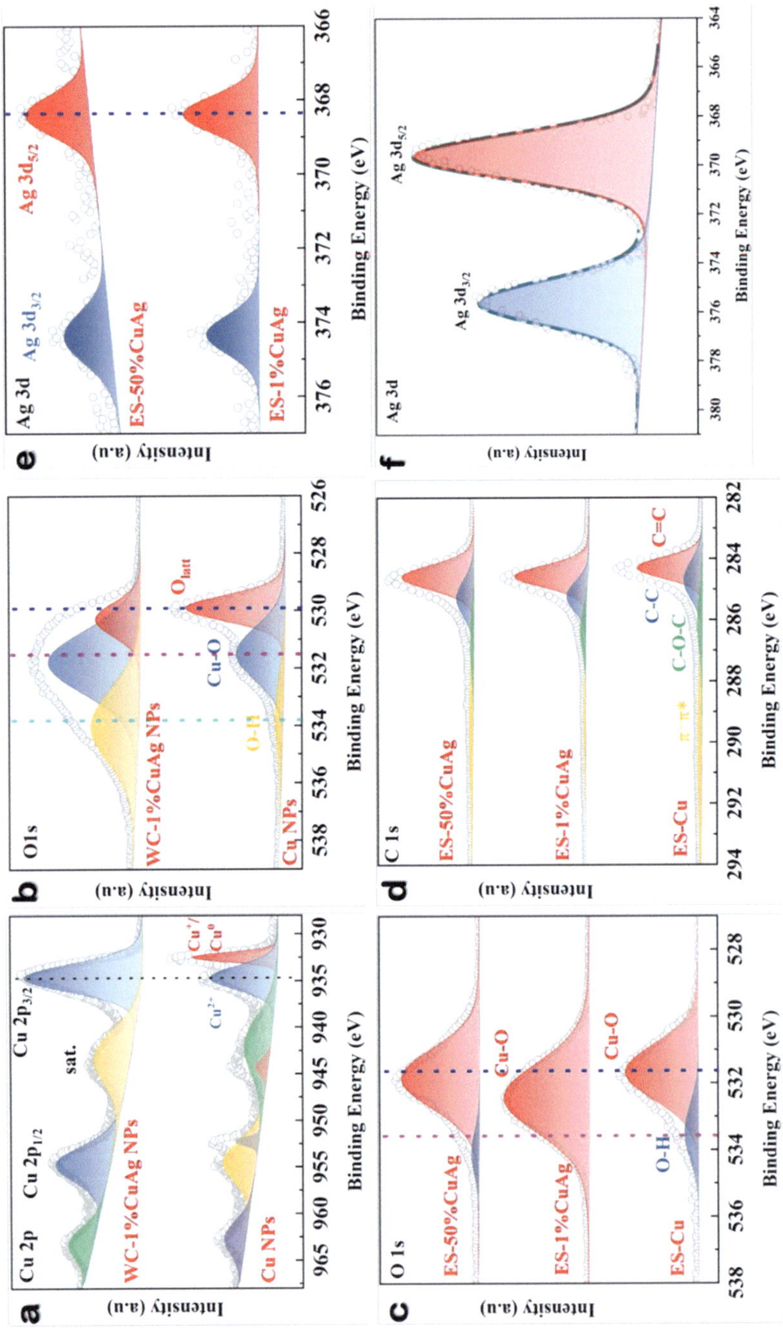

Fig. A.8 **a** Cu 2p XPS spectra and **b** O 1s XPS spectra of Cu NPs and WC-1%CuAg NPs samples; **c** O 1s XPS spectra and **d** C 1s spectra of ES-Cu, ES-1%CuAg, and ES-50%CuAg samples. **e** Ag 3d XPS spectra of ES-1%CuAg and ES-50%CuAg catalysts. **f** Ag 3d XPS spectra of WC-1%CuAg sample

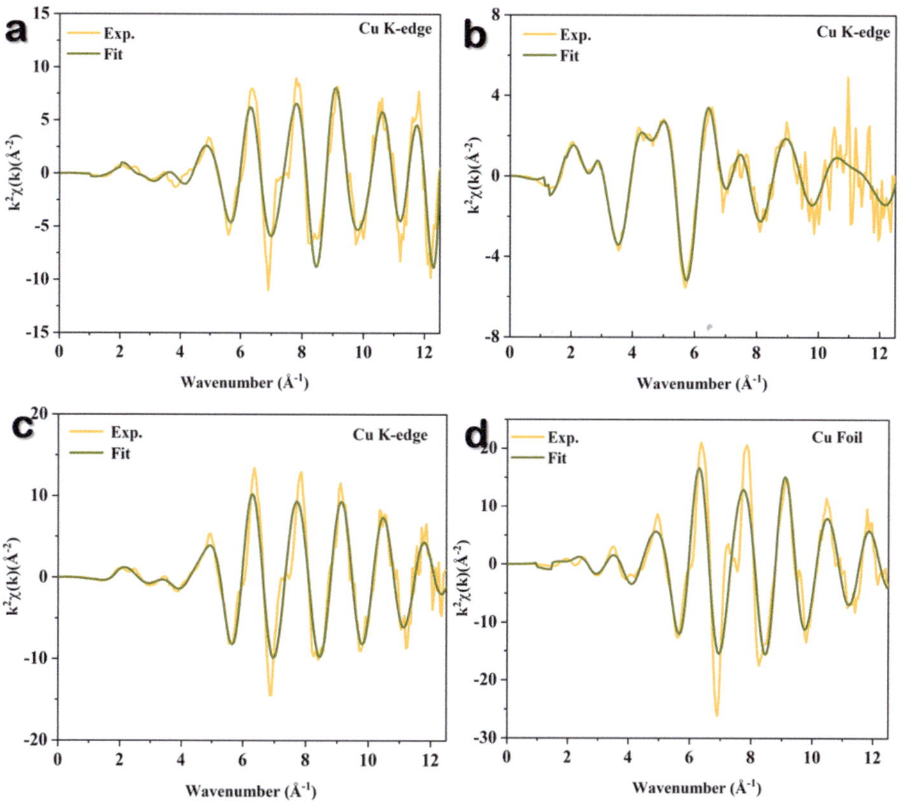

Fig. A.9 Experimental and fitting spectra of Cu K-edge in k space of **a** ES-Cu, **b** ES-1%CuAg, **c** ES-50%CuAg and **d** Cu foil

Table A.2 Fitted EXAFS parameters at Cu and Ag K-edge for ES-Cu, ES-1%CuAg, ES-50%CuAg, and Cu foil

	Bond	Length Å	CN	ΔE	$\delta^2(10^{-3})$	R-factor
Cu foil	Cu–Cu	2.54(0.01)	12	3.6(0.6)	8.6(0.5)	0.01
ES-Cu	Cu–Cu	2.53(0.02)	5.7(0.8)	4.0(0.7)	5.1(0.5)	0.02
ES-1%CuAg	Cu–O	1.87(0.04)	3.7(0.4)	4.8(0.9)	7.1(0.8)	0.006
	Cu–Cu/Ag	2.87(0.05)	5.4(0.8)	10(0.8)	9.8(0.6)	
ES-50%CuAg	Cu–Cu	2.64(0.06)	1.9(0.7)	5.4(0.7)	4.8(0.7)	0.007
	Cu–Ag	2.77(0.07)	4.2(0.6)	9.4(0.8)	5.9(0.9)	
	Cu–Ag	2.86(0.04)	3.2(0.9)	5.6(0.6)	8.4(0.6)	

CN: coordination number; δ^2: mean-square disorder; ΔE: the energy difference. R-factor: goodness of EXFAS fitting.

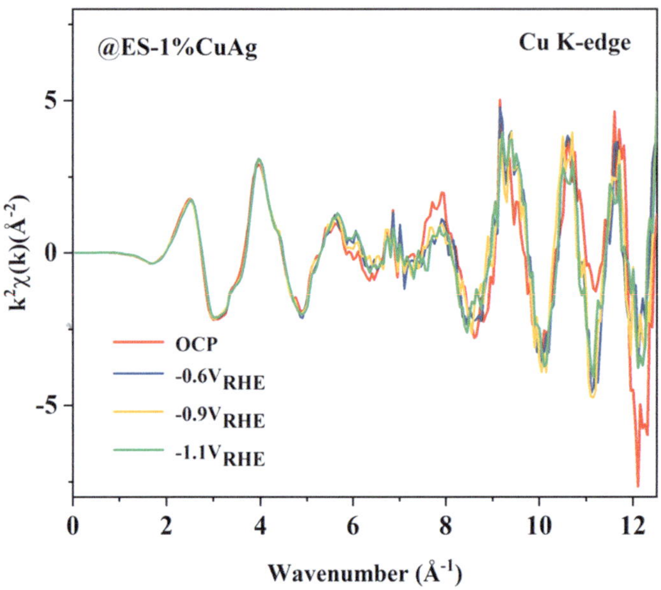

Fig. A.10 In situ XAS spectra analysis of ES-1%CuAg. Experimental k-space spectra of in situ Cu K-edge of ES-1%CuAg under OCP and cathodic potentials from − 0.6 V_{RHE}, − 0.9 V_{RHE} to − 1.1 V_{RHE}

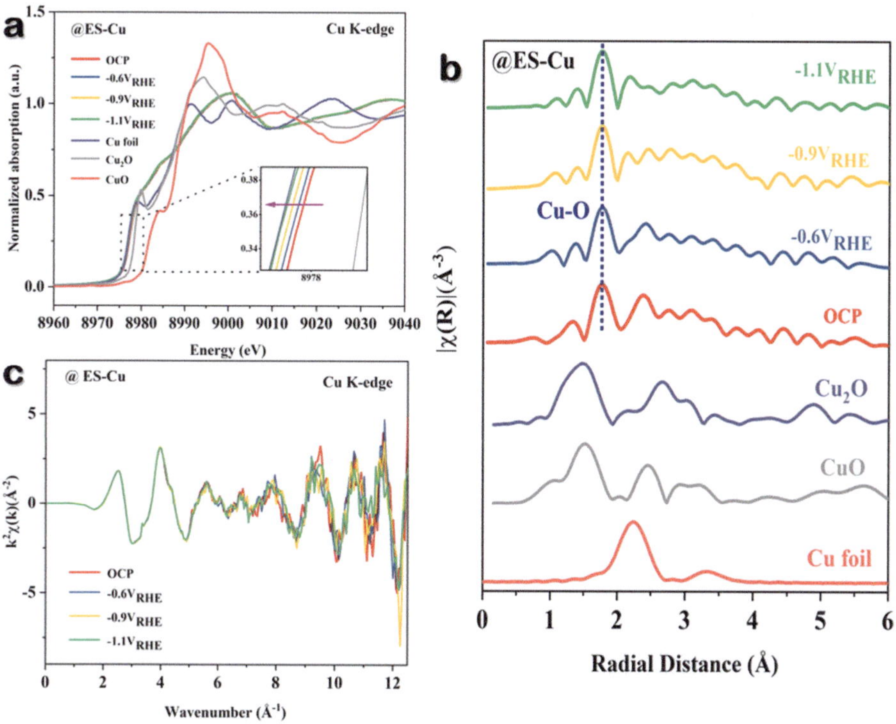

Fig. A.11 **a** In situ Cu K-edge XANES spectra, and **b** in situ Cu K-edge EXAFS spectra, and **c** experimental k-space spectra of in situ Cu K-edge of ES-Cu under OCP and cathodic potentials from − 0.6 V_{RHE}, − 0.9 V_{RHE} to − 1.1 V_{RHE}

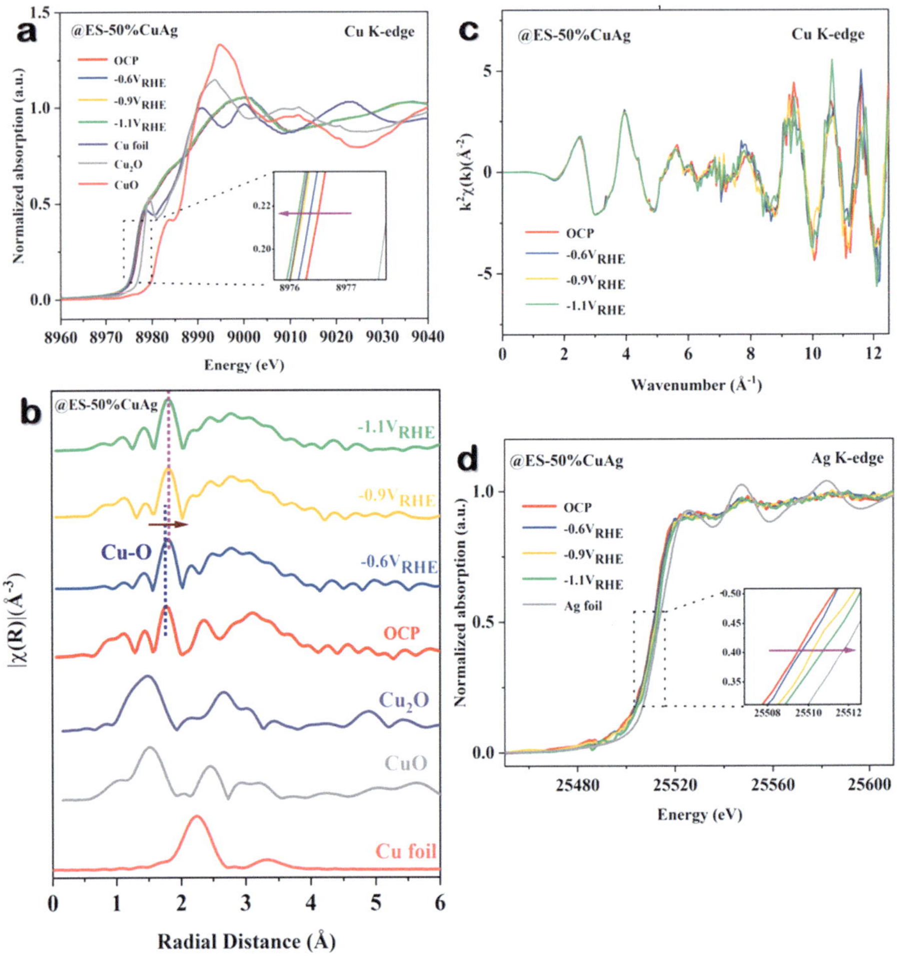

Fig. A.12 **a** In situ Cu K-edge XANES spectra, **b** in situ Cu K-edge EXAFS spectra, **c** experimental k-space spectra, and **d** in situ Ag K-edge XANES spectra of in situ Cu K-edge of ES-50%CuAg under OCP and cathodic potentials from $-0.6\ V_{RHE}$, $-0.9\ V_{RHE}$ to $-1.1\ V_{RHE}$

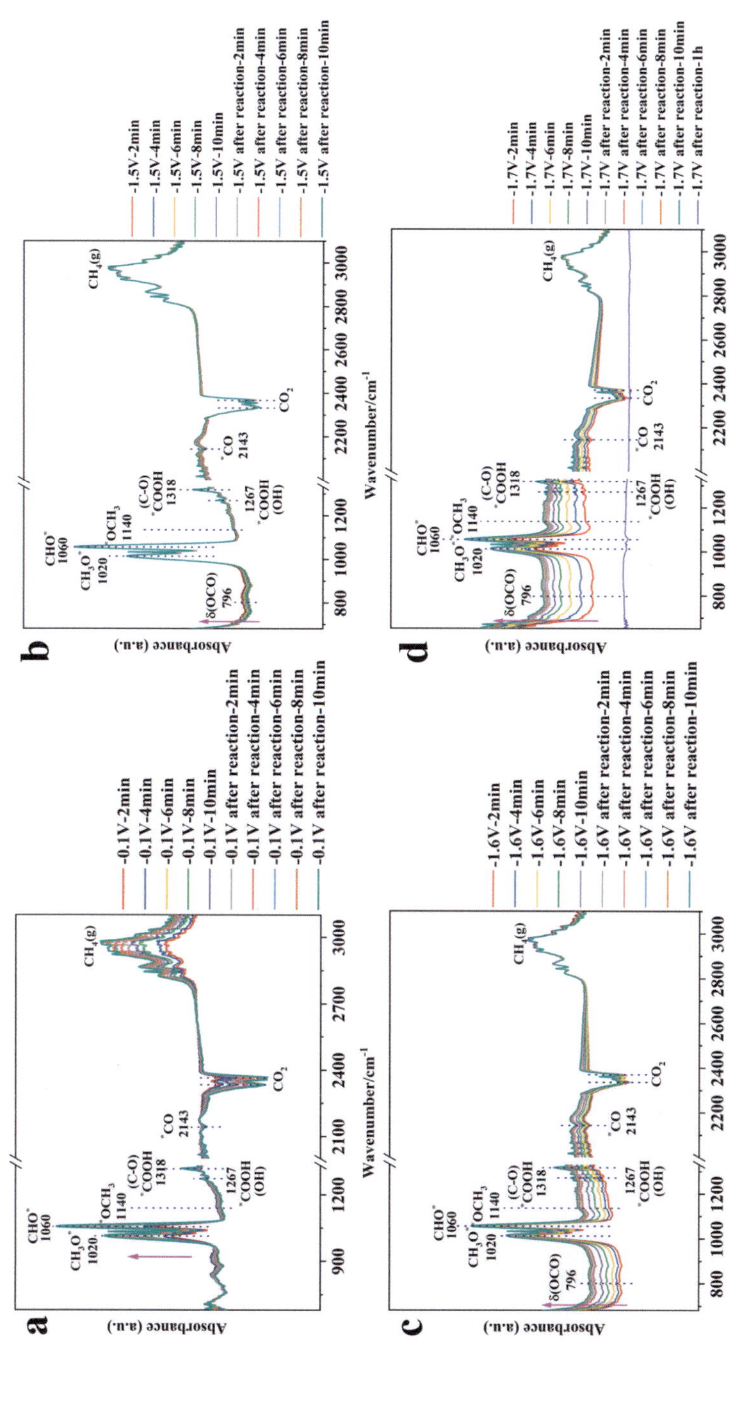

Fig. A.13 In situ ATR FT-IR spectra of ES-1%CuAg sample under an applied potential of **a** – 0.1 V, **b** – 1.5 V, **c** – 1.6 V and **d** – 1.7 V

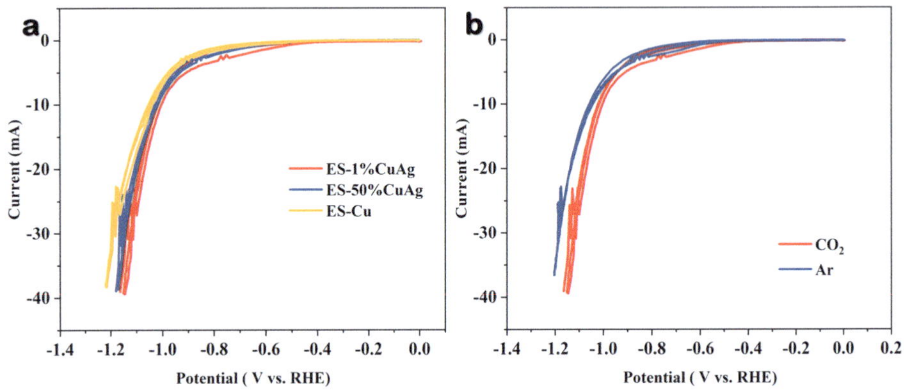

Fig. A.14 **a** Comparison of CV curves of ES-Cu, ES-1%CuAg and ES-50%CuAg samples. **b** Comparison of CV curves for ES-1%CuAg under CO_2-saturated and Ar-saturated 0.1 M $KHCO_3$ electrolyte

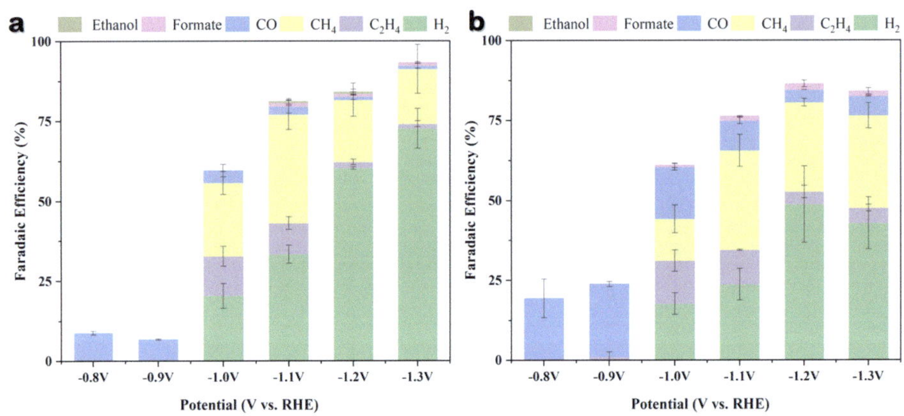

Fig. A.15 Faradaic efficiency of reduced products on **a** ES-Cu and **b** ES-50%CuAg as a function of the electrode potential

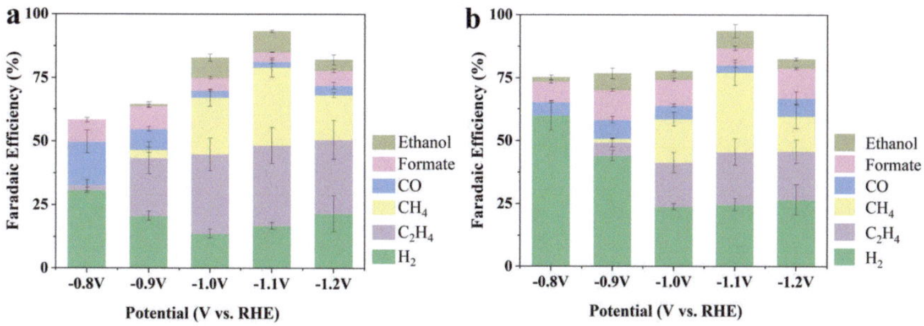

Fig. A.16 Faradaic efficiency of reduced products on **a** commercial Cu NPs and **b** WC-1%CuAg NPs as a function of the electrode potential

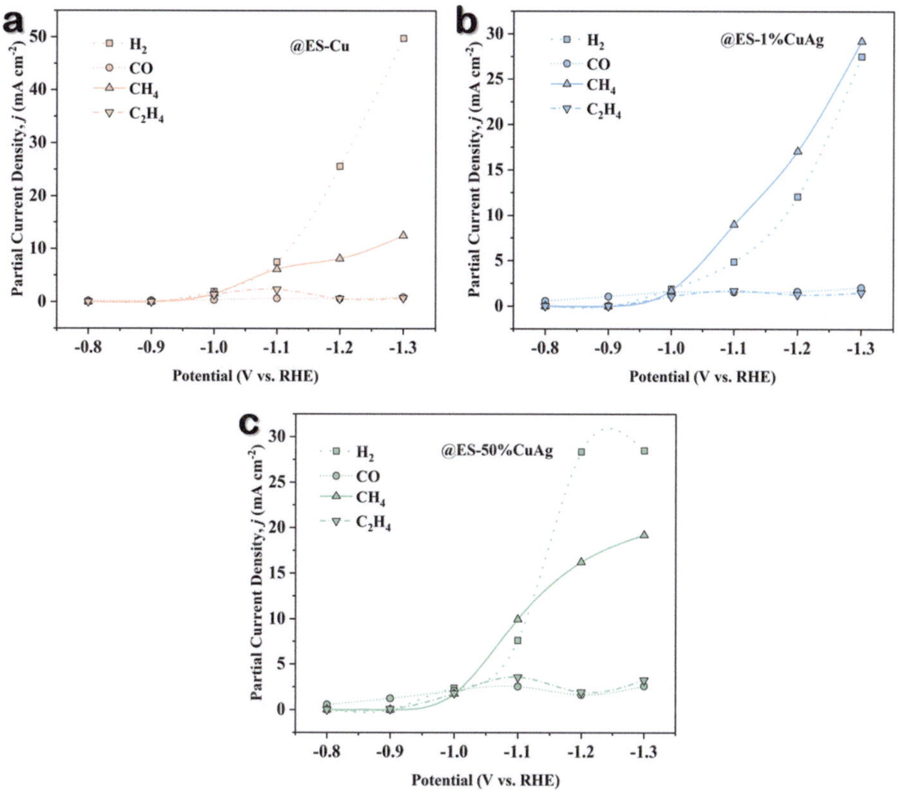

Fig. A.17 The partial current density of reduced products of **a** ES-Cu, **b** ES-1%CuAg and **c** ES-50%CuAg as function of applied negative potentials

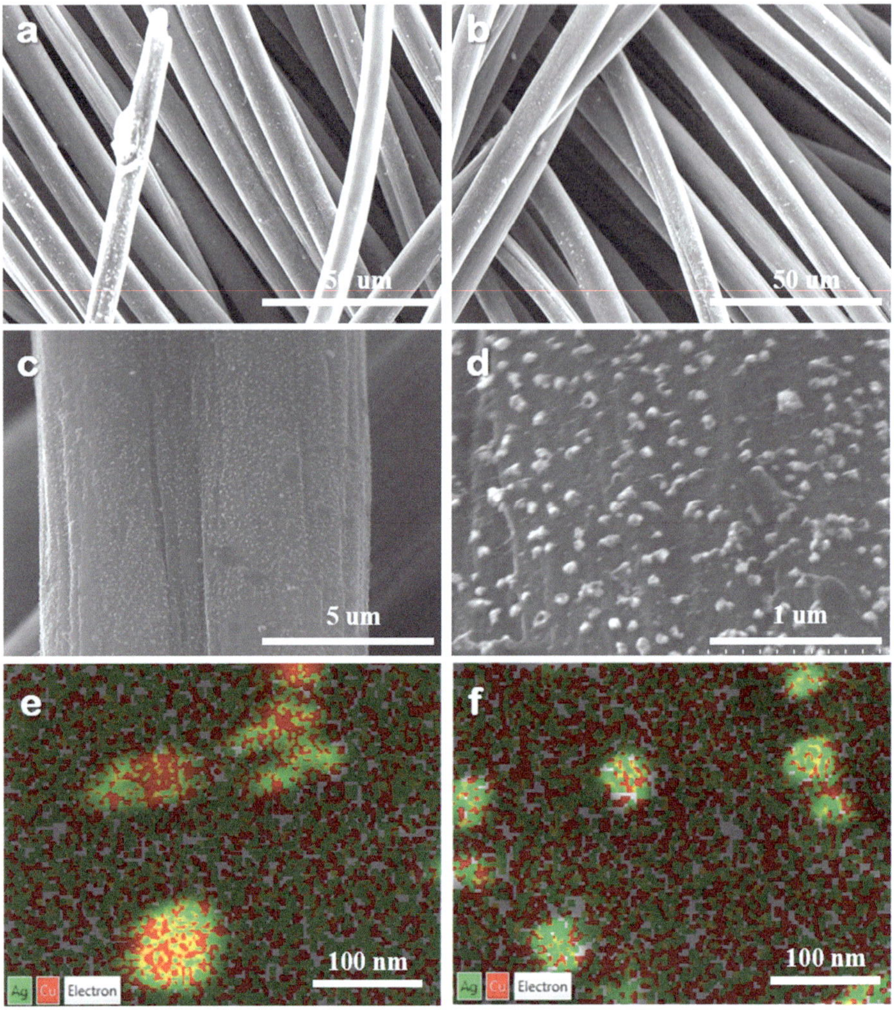

Fig. A.18 **a**–**d** SEM images and **e**, **f** corresponding EDS element mapping of ES-1%CuAg sample after long-term stability test

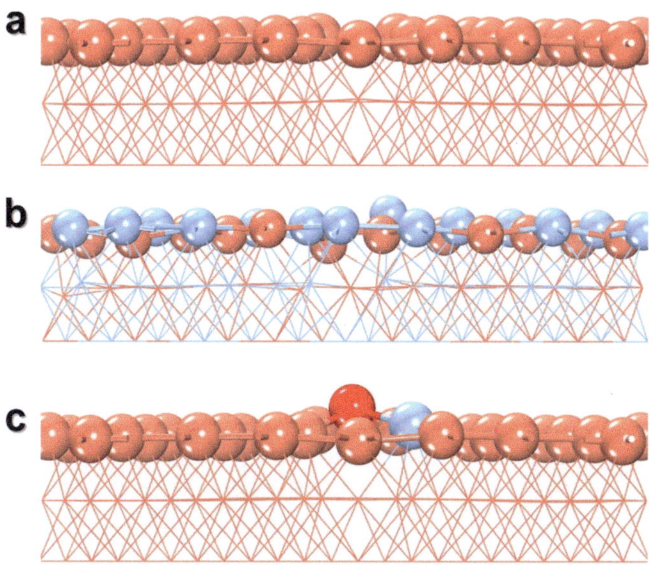

Fig. A.19 Atomic models for DFT. Side views of **a** ES-Cu, **b** ES-50%CuAg, and **c** ES-1%CuAg. The GBs in ES-Cu were simulated by the 3-layer Cu Σ5(210) tilt GB model, and half of Cu atoms in Cu Σ5(210) were replaced by Ag atoms to simulate ES-50%CuAg. For ES-1%CuAg, a surface Cu atom located at GBs was replaced by an Ag atom due to the favorable Ag aggregation at Cu GBs, and an O atom bonding to Cu and Ag atoms was added to model the oxidation effect. The bottom layers in all 3-layer slab models were fixed at the bulk positions during optimizations

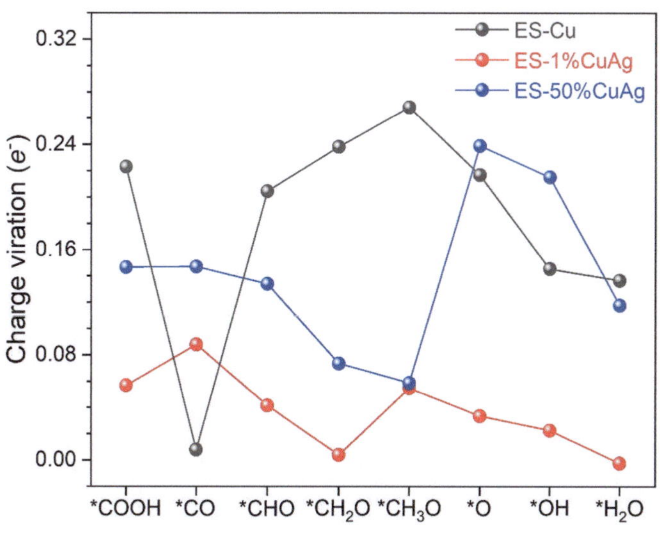

Fig. A.20 Bader charge variations of active Cu sites in ES-Cu, ES-1%CuAg, and ES-50%CuAg relative to the corresponding clean surfaces during ECR

Appendix B: Supporting Information for Chap. 4

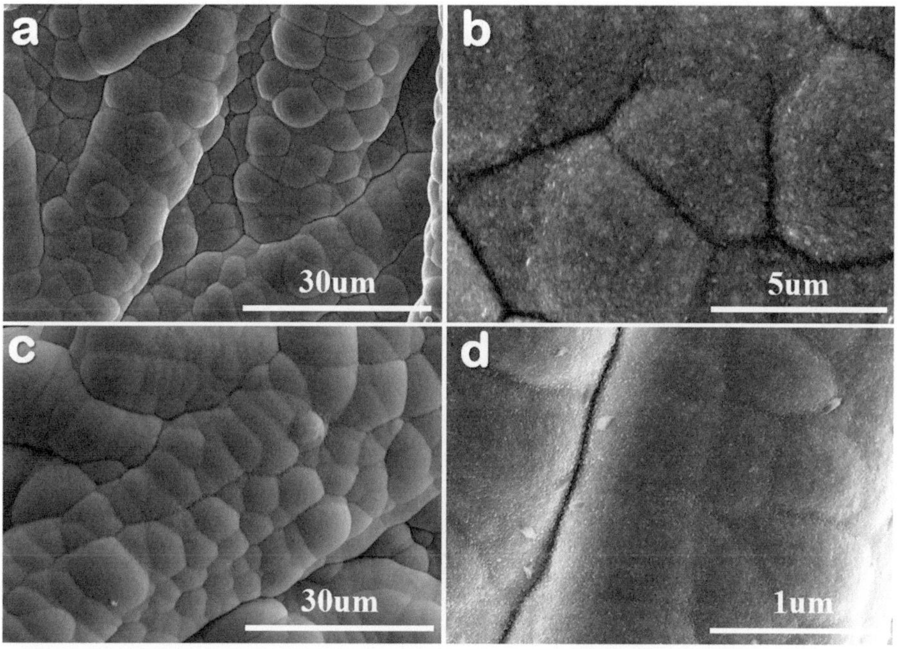

Fig. B.1 Scanning electron microscopy (SEM) images of **a**, **b** Cu-GBs and **c**, **d** Cu-GBOs

L. Wang et al., *Modulation Strategies of Cu-based Electrocatalysts for Enhancing Electrocatalytic CO_2 Conversion*, Synthesis Lectures on Green Energy and Technology,
https://doi.org/10.1007/978-3-031-98056-5

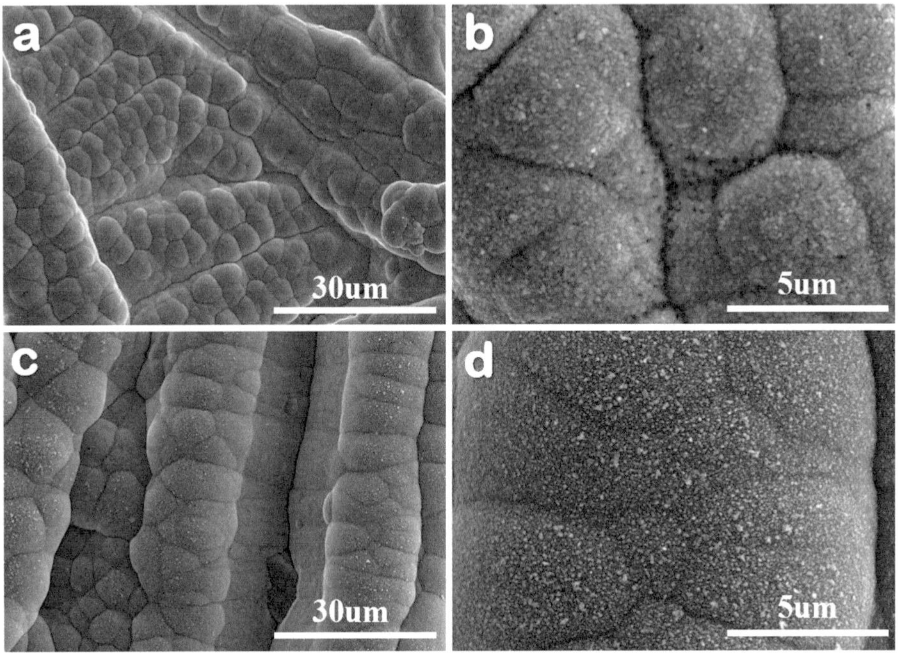

Fig. B.2 SEM images of **a**, **b** Ag/Cu-GBO and **c**, **d** CuAg-GBs

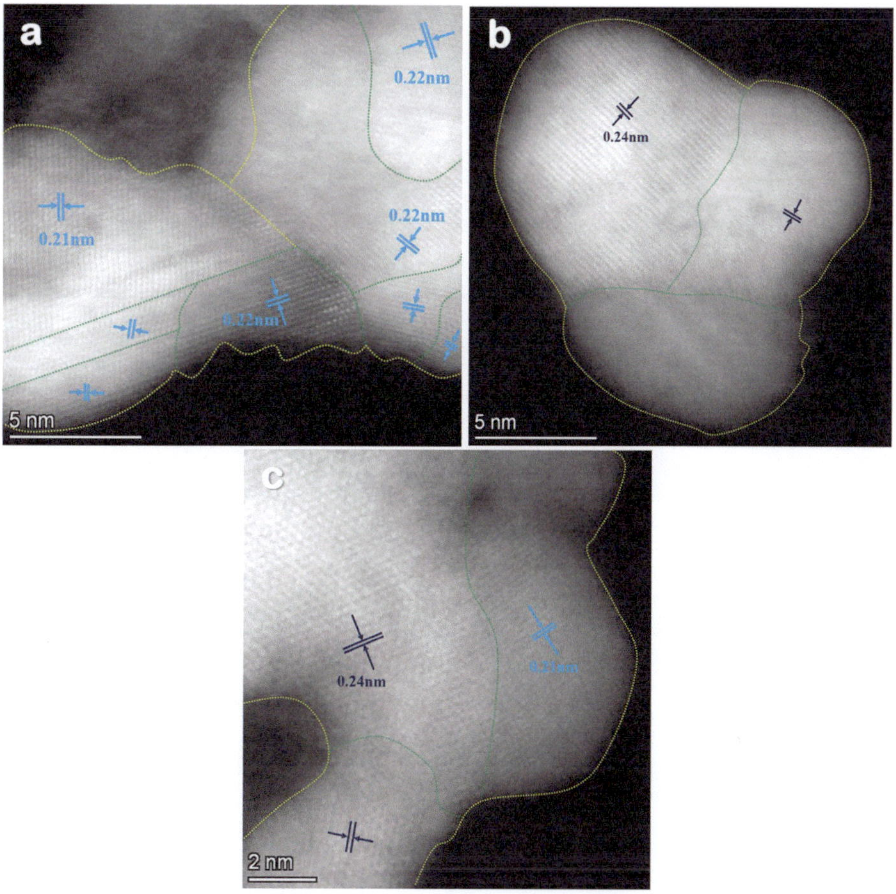

Fig. B.3 AC-HAADF-STEM images of **a** Ag/Cu-GBO and **b**, **c** CuAg-GBs

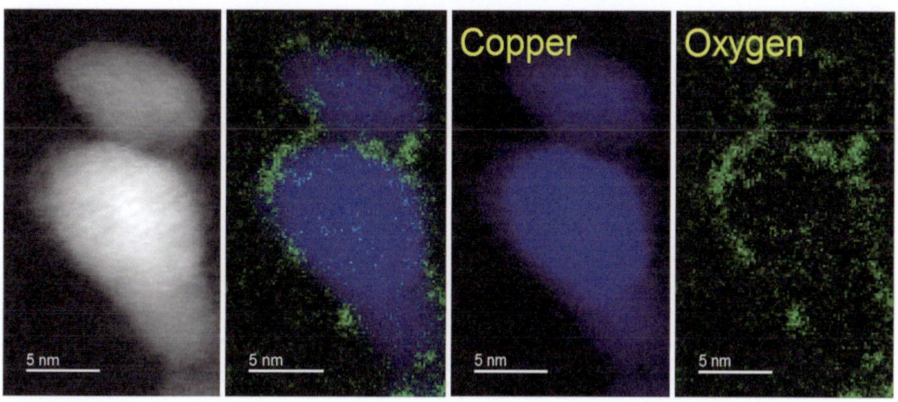

Fig. B.4 Electron energy loss spectroscopy (EELS) maps of Cu-GBO catalyst

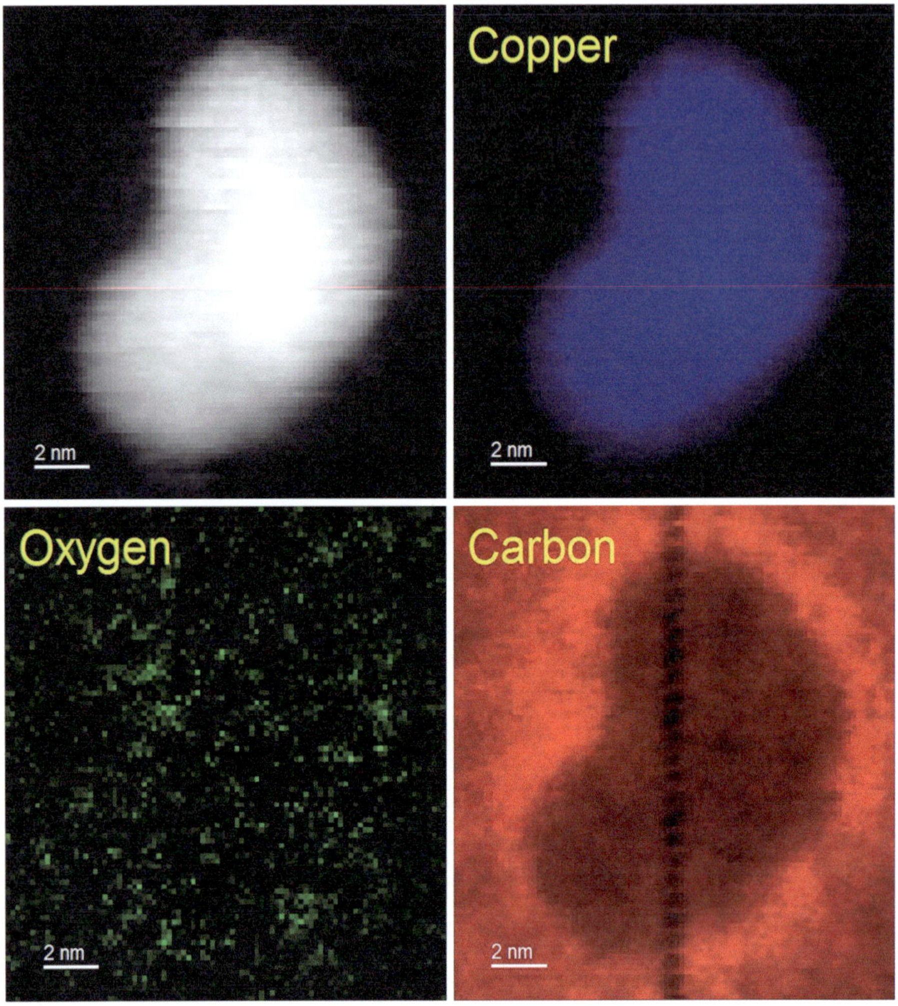

Fig. B.5 Electron energy loss spectroscopy (EELS) maps of Cu-GBs

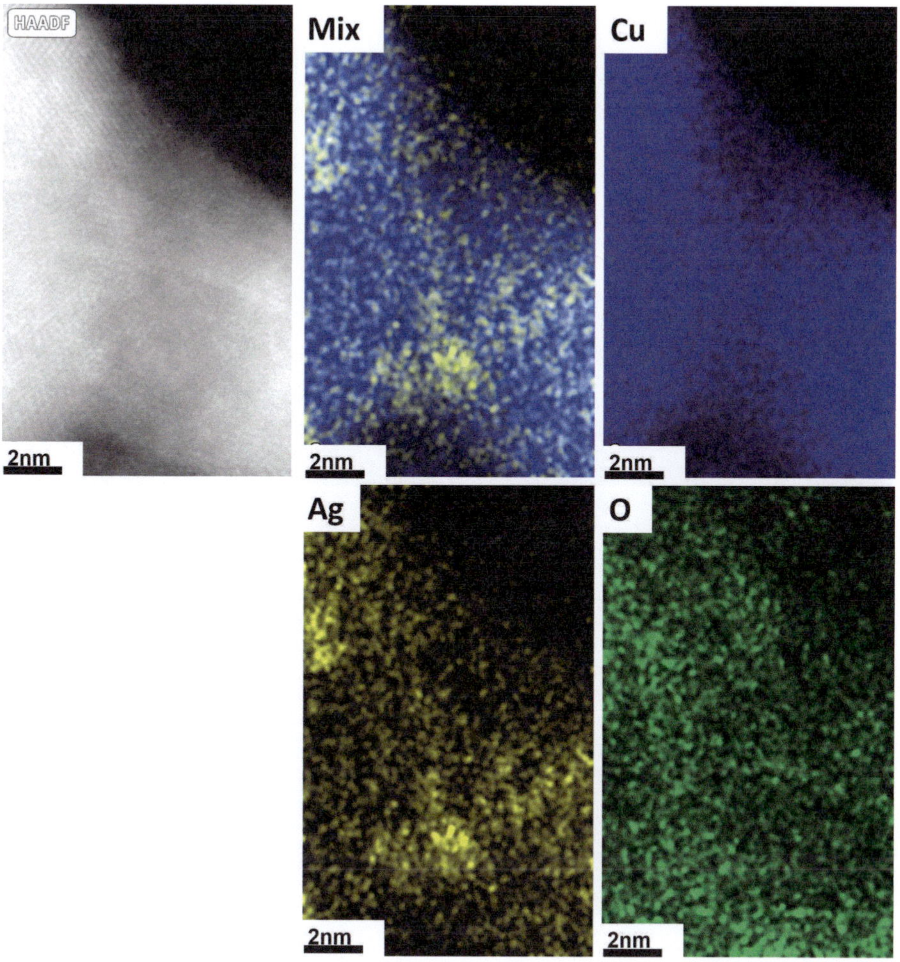

Fig. B.6 STEM-EDS elemental mapping of Ag/Cu-GBO

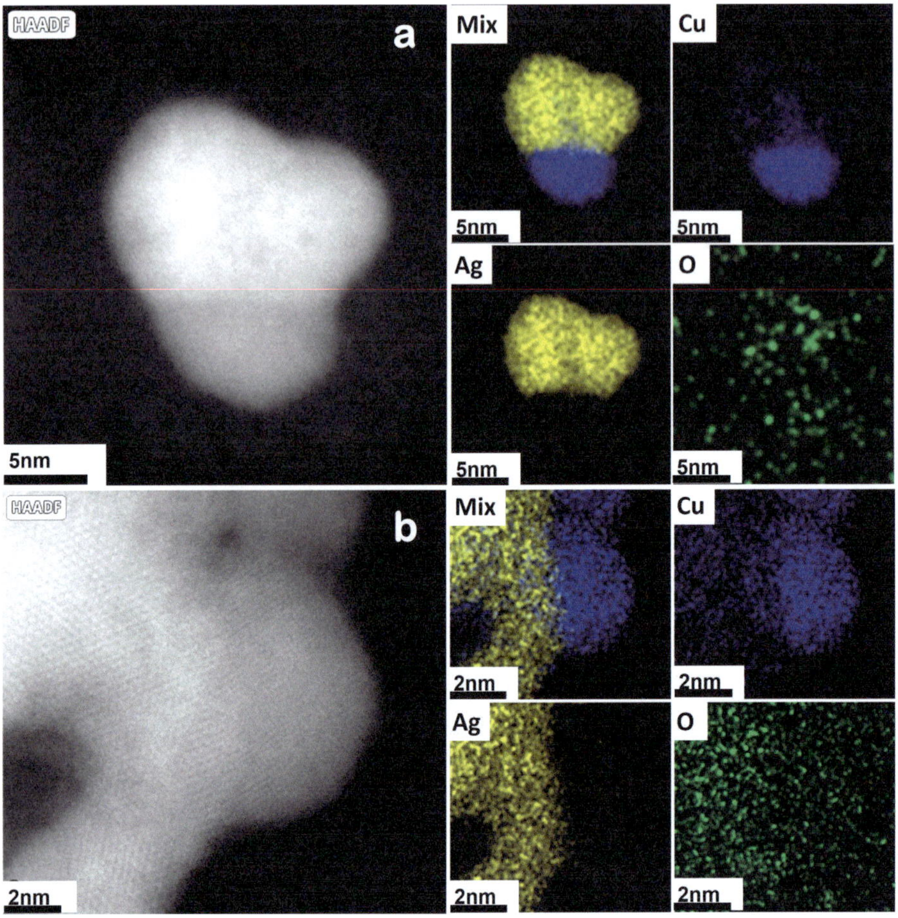

Fig. B.7 **a**, **b** STEM-EDS maps of CuAg-GBs catalysts with the corresponding HAADF-STEM images

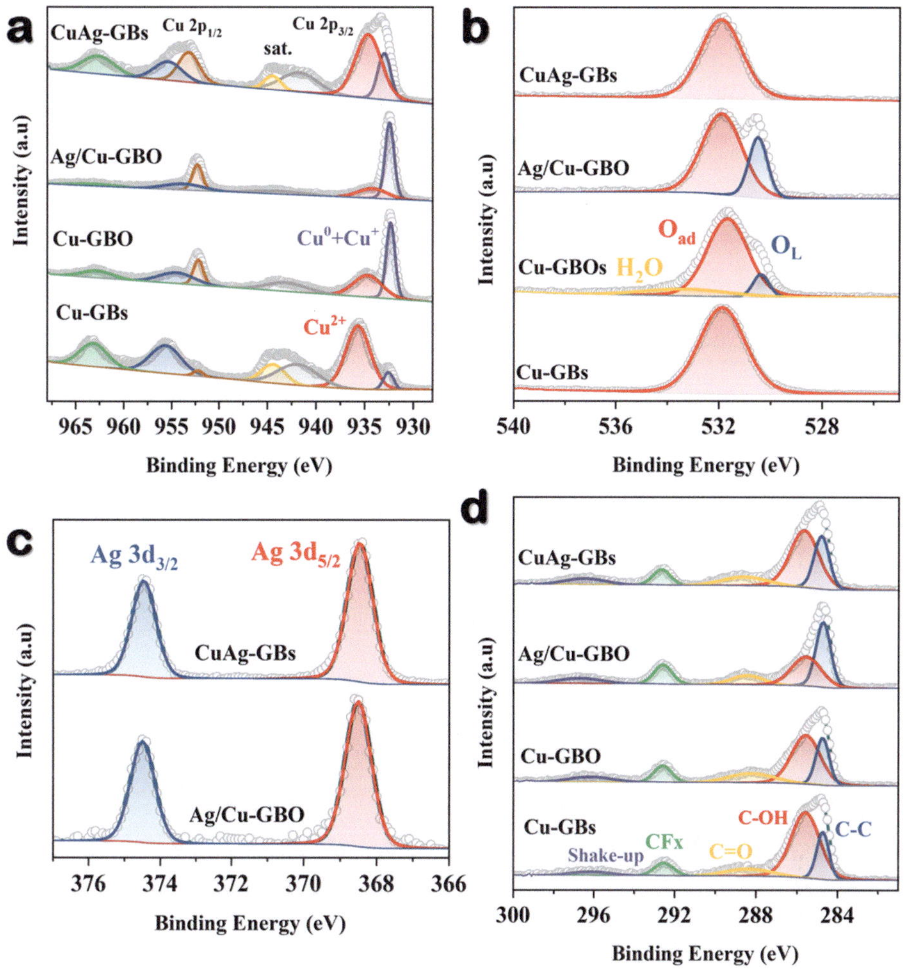

Fig. B.8 **a** Cu 2p XPS spectra, **b** O 1s XPS spectra, **c** Ag 3d XPS spectra, and **d** C 1s XPS spectra of Cu-GBs, Cu-GBO, Ag/Cu-GBO, and CuAg-GBs catalysts

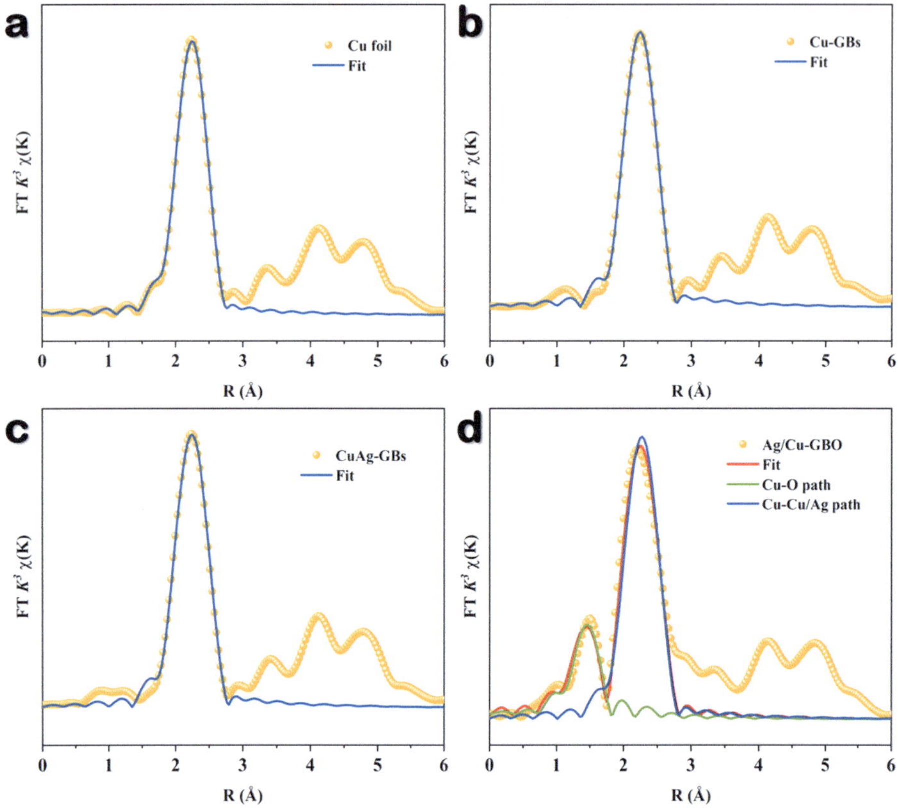

Fig. B.9 Cu K-edge EXAFS experimental and fitting spectrum of **a** Cu foil, **b** Cu-GBs, **c** CuAg-GBs, and **d** Ag/Cu-GBO catalysts

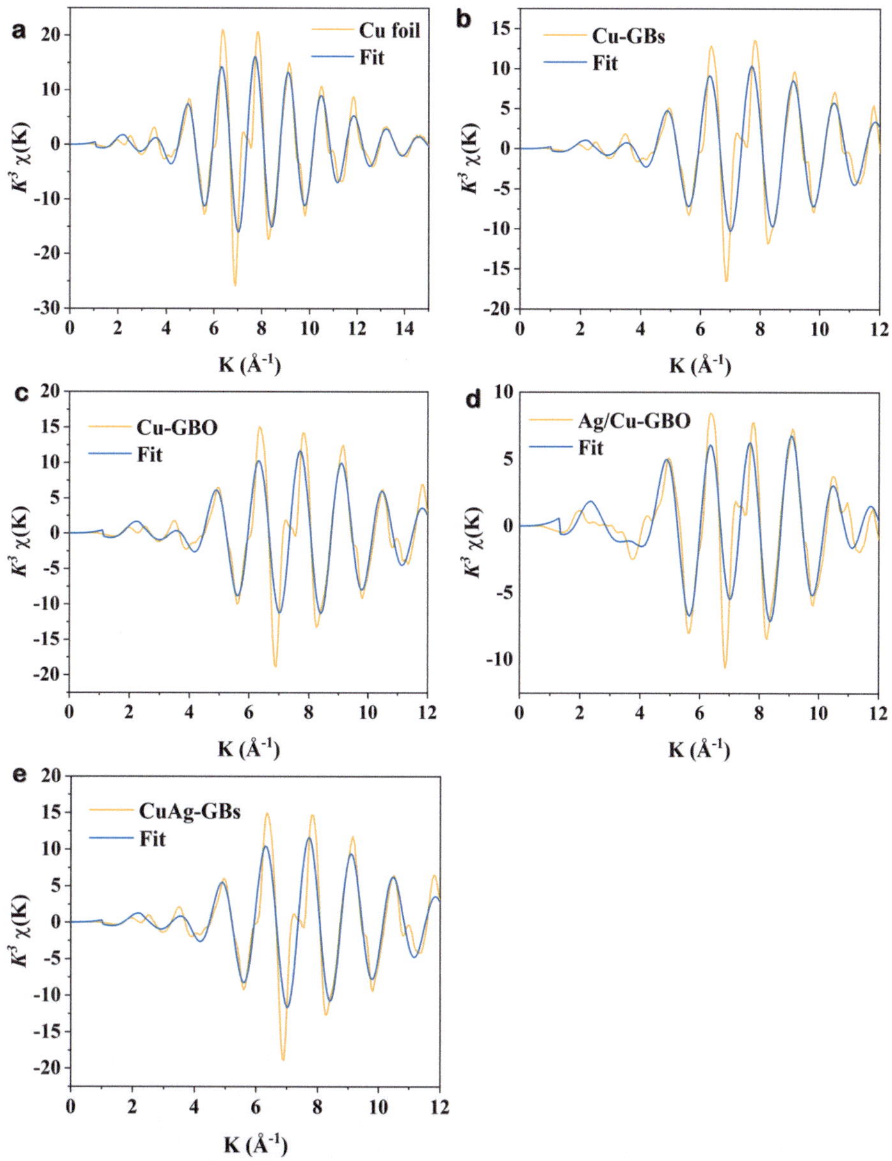

Fig. B.10 Experimental and fitting spectra of Cu K-edge in k space of **a** Cu foil, **b** Cu-GBs, **c** Cu-GBO, **d** Ag/Cu-GBO, and **e** CuAg-GBs catalysts

Table B.1 EXAFS fitting parameters at the Cu K-edge for various samples ($S_0^2 = 0.90$ from Cu-foil)

	Shell	CN[a]	R[b](Å)	σ^{2c}(Å^2)	ΔE_0[d](eV)	R factor
Cu-foil	Cu–Cu	12	2.54 ± 0.01	0.0088	4.5 ± 0.5	0.0027
Cu-GBs	Cu–Cu	7.7 ± 0.2	2.54 ± 0.01	0.0088	3.8 ± 0.8	0.0059
Cu-GBO	Cu–O	0.4 ± 0.2	1.81 ± 0.01	0.0030	5.0 ± 1.0	0.0099
	Cu–Cu	9.2 ± 0.3	2.54 ± 0.01	0.0092		
Ag/Cu-GBO	Cu–O	1.1 ± 0.2	1.84 ± 0.02	0.0030	6.5 ± 1.9	0.0193
	Cu–Cu/Ag	6.0 ± 0.4	2.55 ± 0.01	0.0100		
CuAg-GBs	Cu–Cu/Ag	9.0 ± 0.2	2.54 ± 0.01	0.0092	4.2 ± 0.8	0.0064

[a]*CN*: coordination numbers; [b]*R*: bond distance; [c]σ^2: Debye–Waller factors; [d]ΔE_0: the inner potential correction. R factor: goodness of fit. Error bounds that characterize the structural parameters obtained by EXAFS spectroscopy were estimated as CN ± 20%; R ± 1%; σ^2 ± 20%

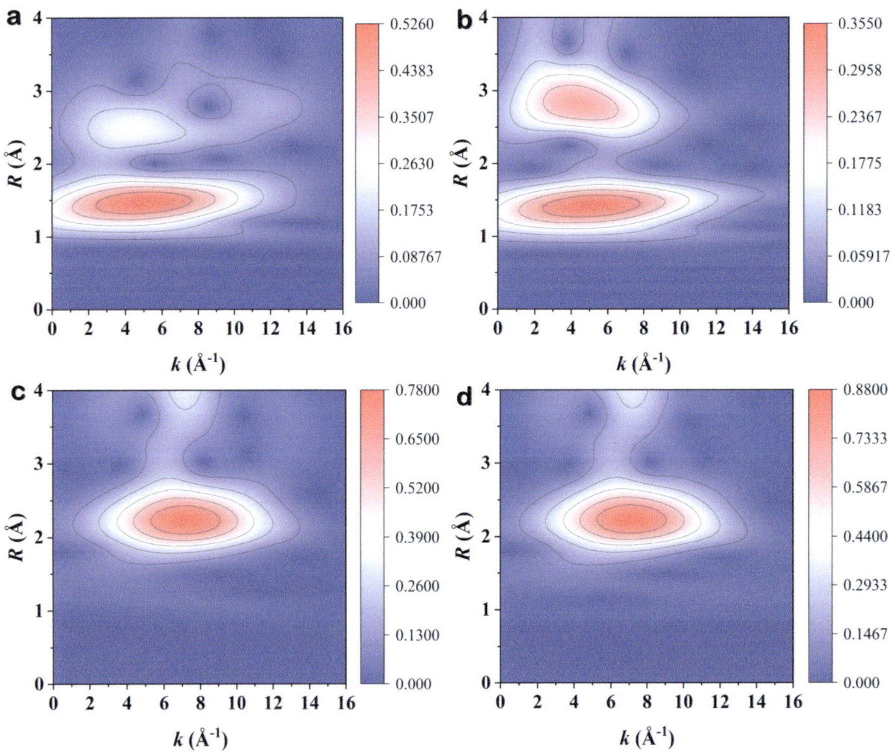

Fig. B.11 Wavelet transforms for the k^3-weighted EXAFS signals of **a** CuO reference, **b** Cu_2O reference, **c** Cu-GBs, and **d** CuAg-GBs

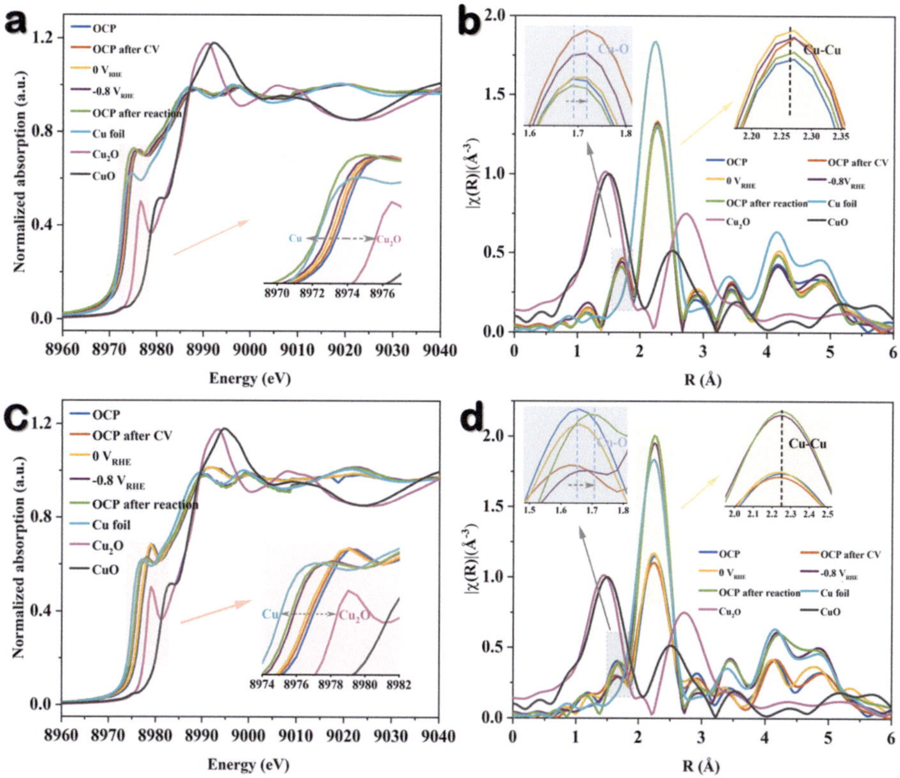

Fig. B.12 **a** In situ Cu K-edge XANES spectra and **b** in situ Cu K-edge EXAFS spectra of Cu-GBs at different applied potentials (including open circuit potential (OCP), OCP after cyclic voltammetry (CV), applied cathodic potentials (0 V_{RHE} and − 0.8 V_{RHE}), and OCP after reaction); **c** in situ Cu K-edge XANES spectra and **d** in situ Cu K-edge EXAFS spectra of Ag/Cu-GBO at different applied potentials (including open circuit potential (OCP), OCP after cyclic voltammetry (CV), applied cathodic potentials (0 V_{RHE} and − 0.8 V_{RHE}), and OCP after reaction)

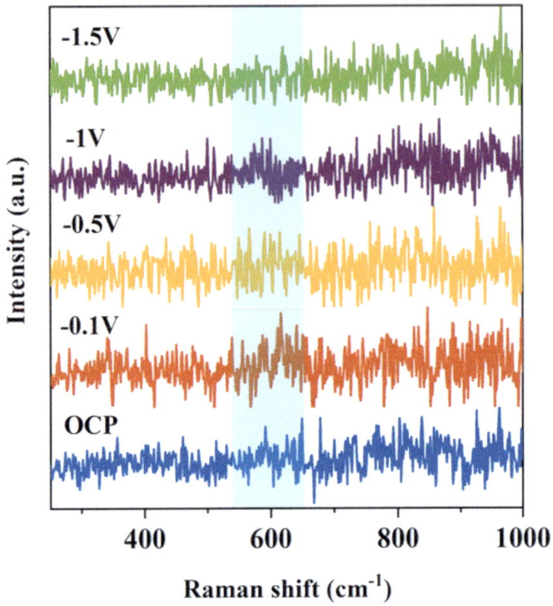

Fig. B.13 In situ Raman spectra collected from Cu-GBs catalyst at different applied potentials (OCP, − 0.1 V, − 0.5 V, − 1 V, and − 1.5 V)

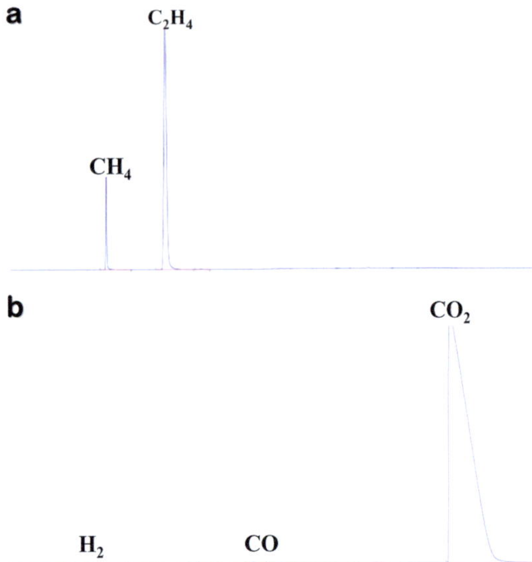

Fig. B.14 **a**, **b** Representative GC trace of gaseous product produced during ECR for Cu-GBO at − 1.0 V_{RHE} in H-cell using neutral electrolyte

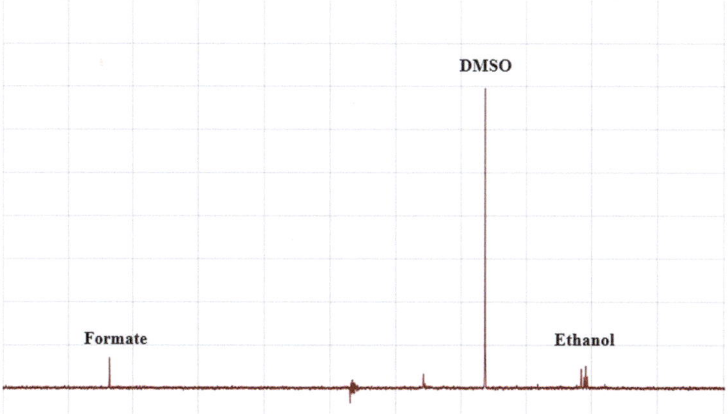

Fig. B.15 Representative ^{1}H-NMR spectrum for catholyte following ECR on Cu-GBO at − 1.0 V_{RHE} in H-cell using neutral electrolyte. DMSO is used as internal standard for ethanol and formate

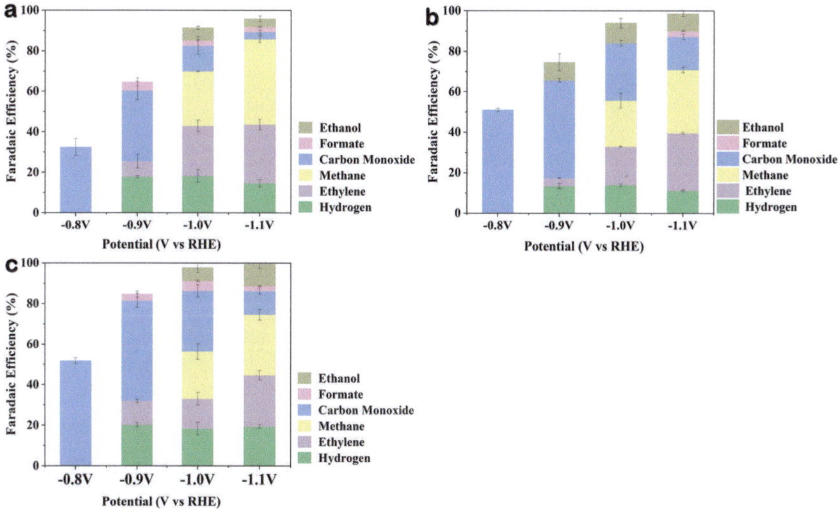

Fig. B.16 Faradaic efficiency of reduced products on **a** Cu-GBs, **b** Ag/Cu-GBO and **c** CuAg-GBs catalysts as a function of the electrode potential in H-cell, respectively

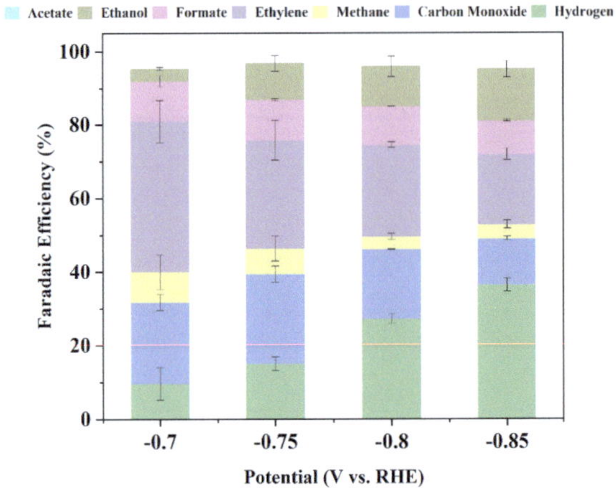

Fig. B.17 Faradaic efficiency of reduced products on Cu-GBO under applied potentials in flow cell using neutral conditions

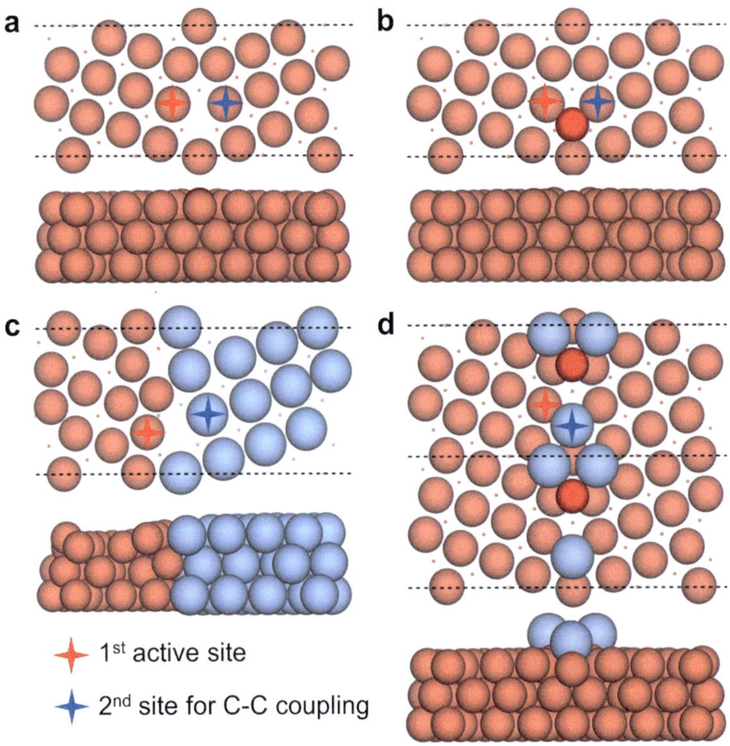

Fig. B.18 Top and side views of **a** Cu-GBs, **b** Cu-GBO, **c** CuAg-GBs, and **d** Ag/Cu-GBO. $*CO_2 \rightarrow *CHO$ proceeds on the 1st active site, and the 2nd active site gets involved in the C–C coupling and the subsequent hydrogenation to ethylene. Color code: Ag, light blue

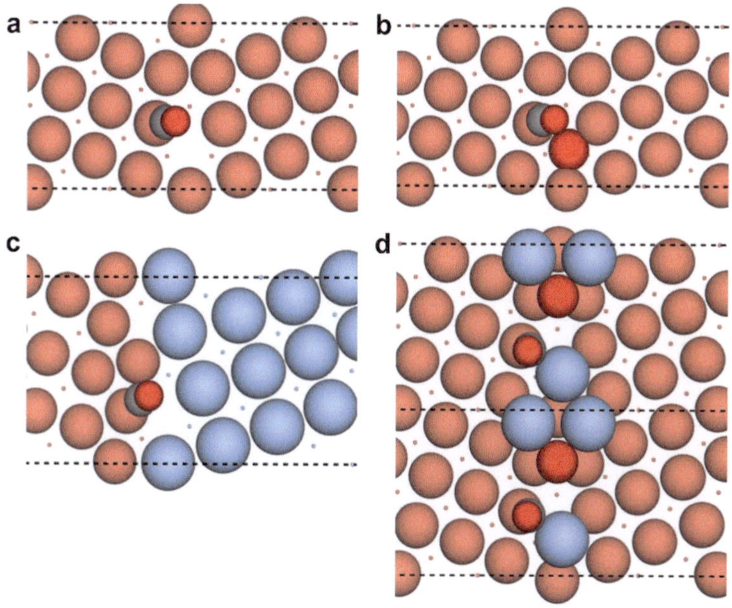

Fig. B.19 Top views of *CO on **a** Cu-GBs, **b** Cu-GBO, **c** CuAg-GBs, and **d** Ag/Cu-GBO

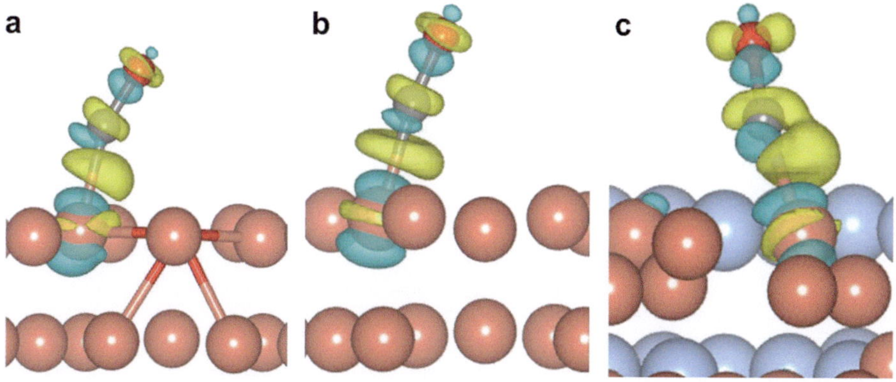

Fig. B.20 3D displays of charge density differences for *CO on **a** Cu-GBO, **b** Cu-GBs, and **c** CuAg-GBs, where yellow and cyan regions denote electron accumulation and depletion, respectively, with an isovalue of 0.005 e^{-}/Å

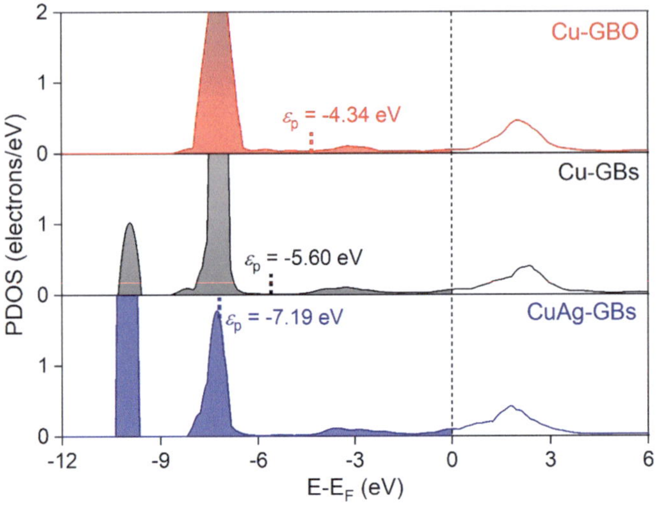

Fig. B.21 PDOS of p orbitals of oxygen atoms of *CO on Cu-GBO, Cu-GBs, and CuAg-GBs

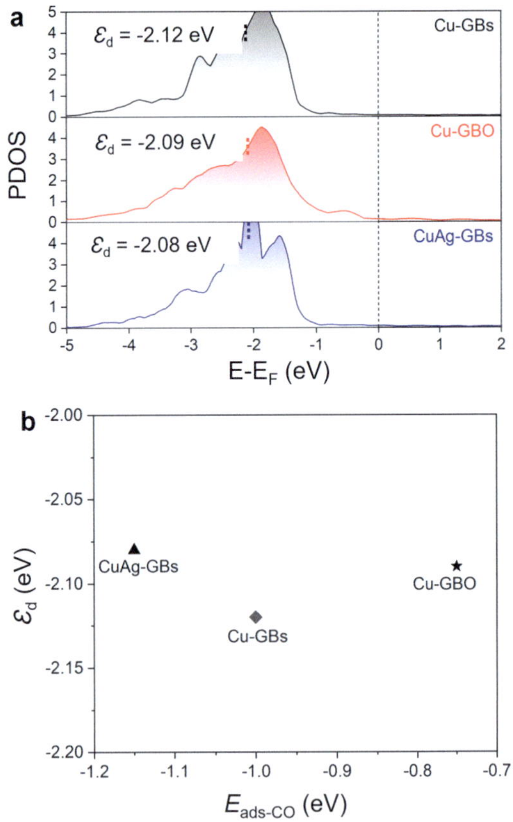

Fig. B.22 **a** PDOS of d orbitals of 1st Cu active site and **b** ε_d and $E_{ads\text{-}CO}$ values of Cu-GBs, Cu-GBO, and CuAg-GBs

Appendix C: Supporting Information for Chap. 5

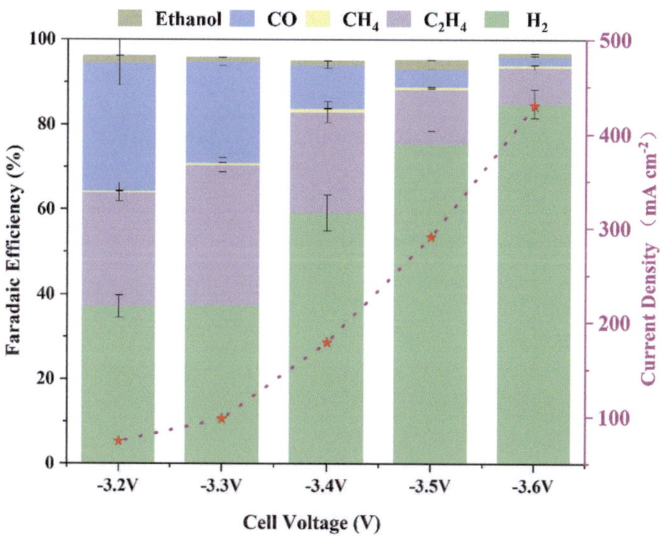

Fig. C.1 ECR performance of Cu NPs in the MEA reactor. All applied potentials were not iR corrected ($R = 0.25 \pm 0.01\ \Omega$). Error bars represent the standard deviations calculated from three independent measurements

L. Wang et al., *Modulation Strategies of Cu-based Electrocatalysts for Enhancing Electrocatalytic CO_2 Conversion*, Synthesis Lectures on Green Energy and Technology,
https://doi.org/10.1007/978-3-031-98056-5

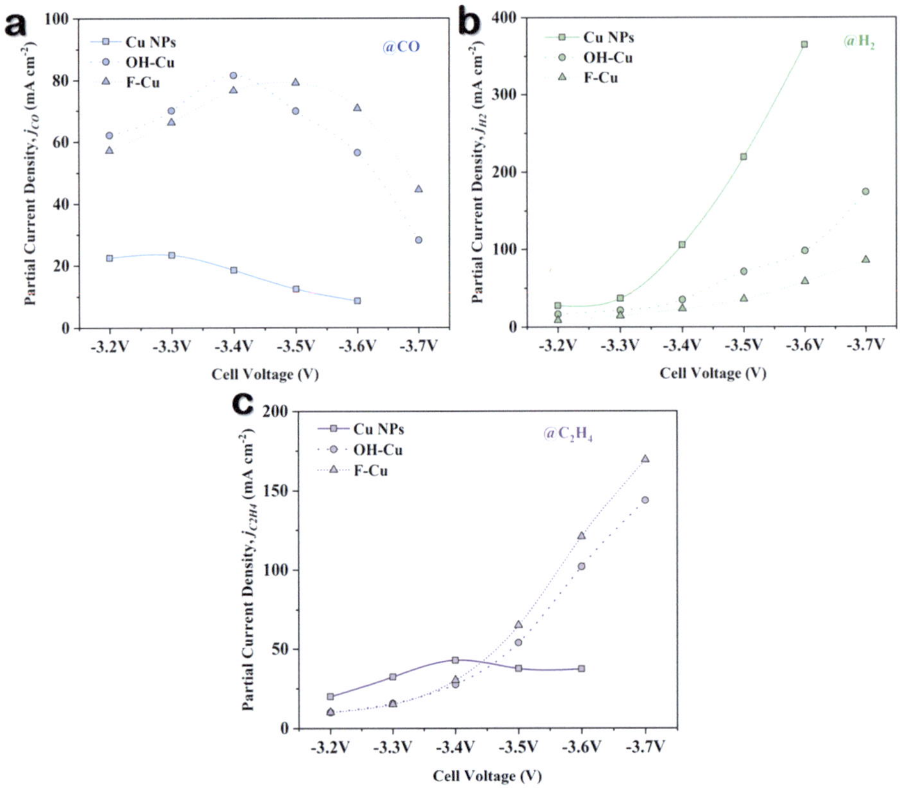

Fig. C.2 The partial current densities towards the **a** CO, **b** H_2, and **c** C_2H_4 products. All applied potentials were not iR corrected (R = 0.25 ± 0.01 Ω)

Fig. C.3 Charging current density against scan rates over for all samples

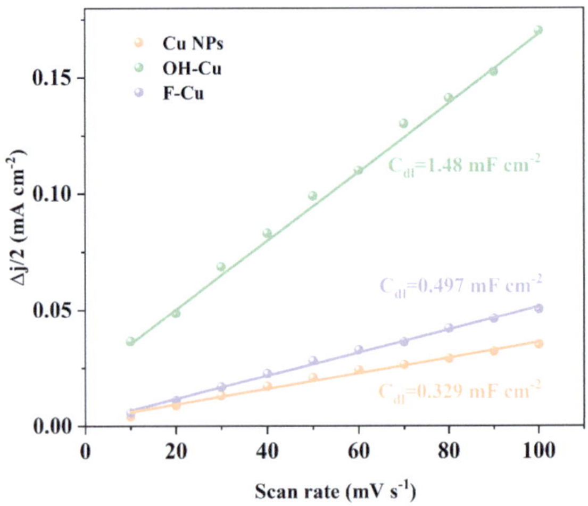

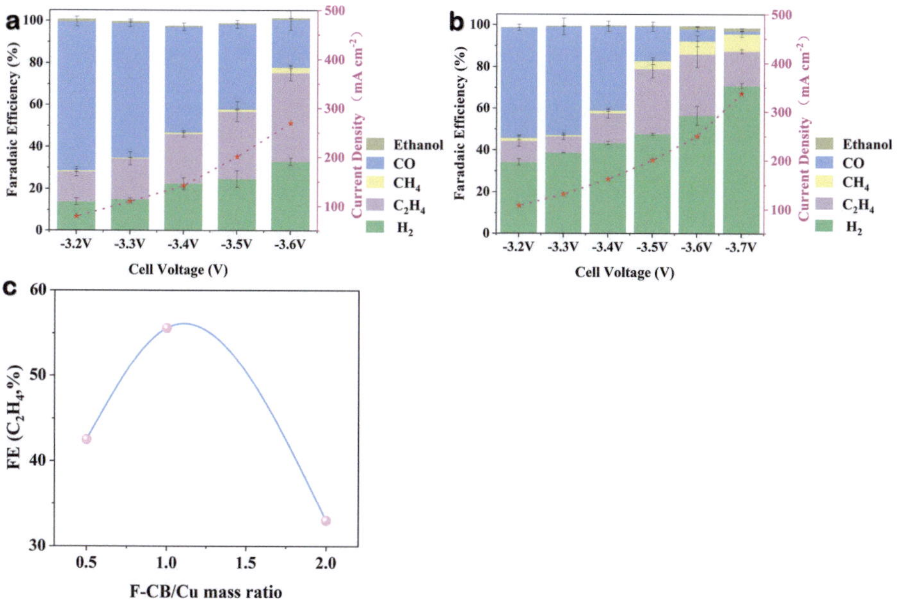

Fig. C.4 ECR performance of F–Cu catalysts with **a** F-CB/Cu = 1:2 and **b** 2:1 in a MEA reactor. **c** The relationship between the selectivity of ethylene and loading of F-CB, where the loading of F-CB corresponds to the interfacial microenvironment of Cu NPs. All applied potentials were not iR corrected (R = 0.25 ± 0.01 Ω). Error bars represent the standard deviations calculated from three independent measurements

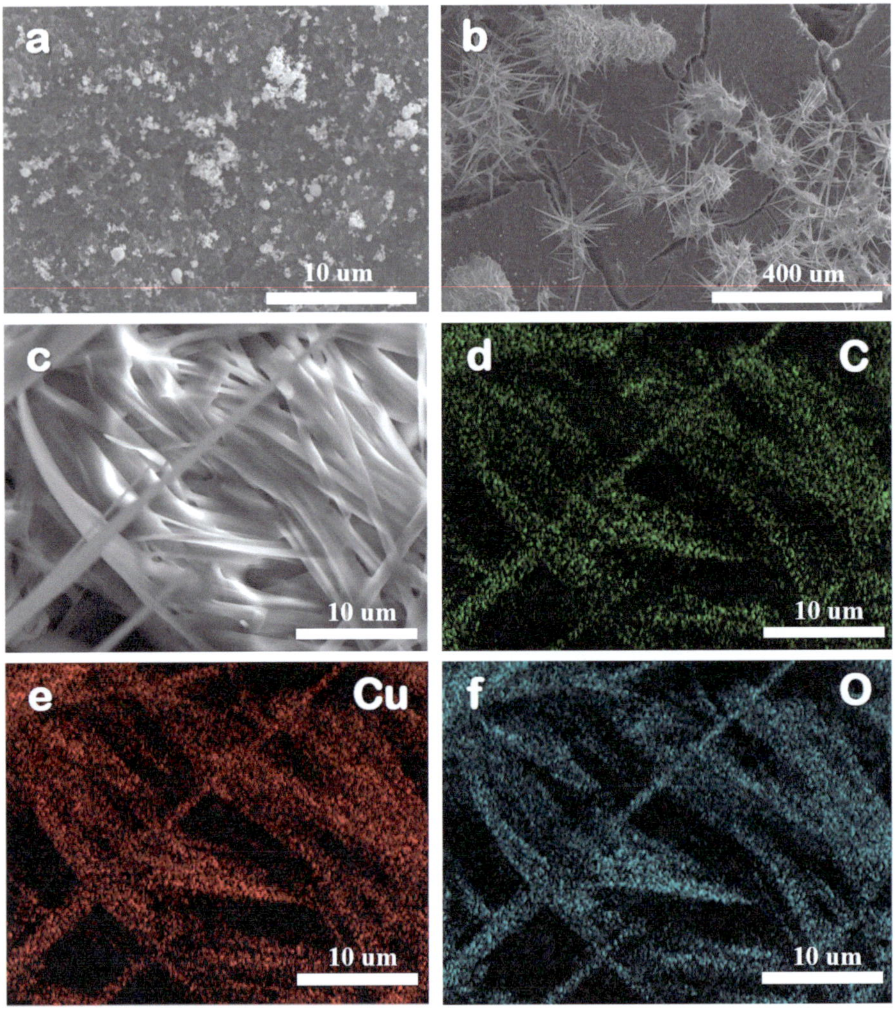

Fig. C.5 SEM images of **a** fresh Cu NPs on the GDE and **b**, **c** used Cu NPs on the GDE. **d**–**f** The EDX mapping images of used Cu NPs on the GDE

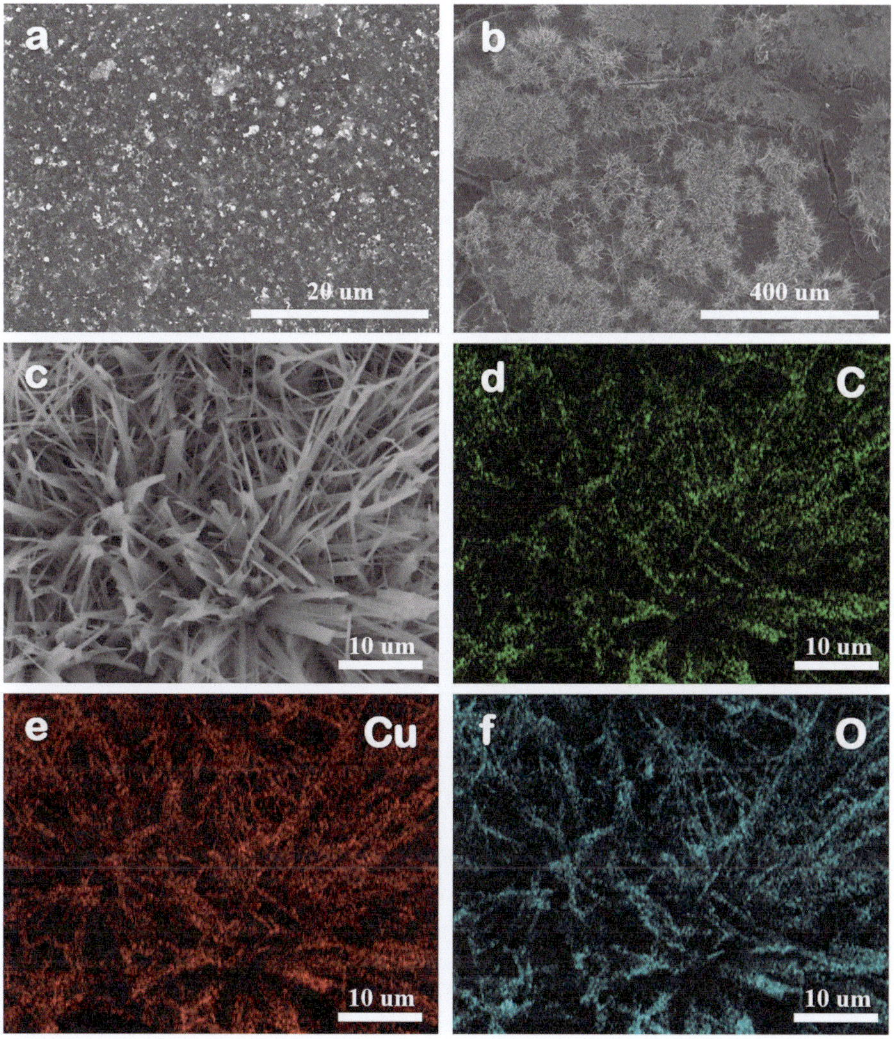

Fig. C.6 SEM images of **a** fresh OH–Cu on the GDE and **b**, **c** used OH–Cu on the GDE. **d–f** The EDX mapping images of used OH–Cu on the GDE

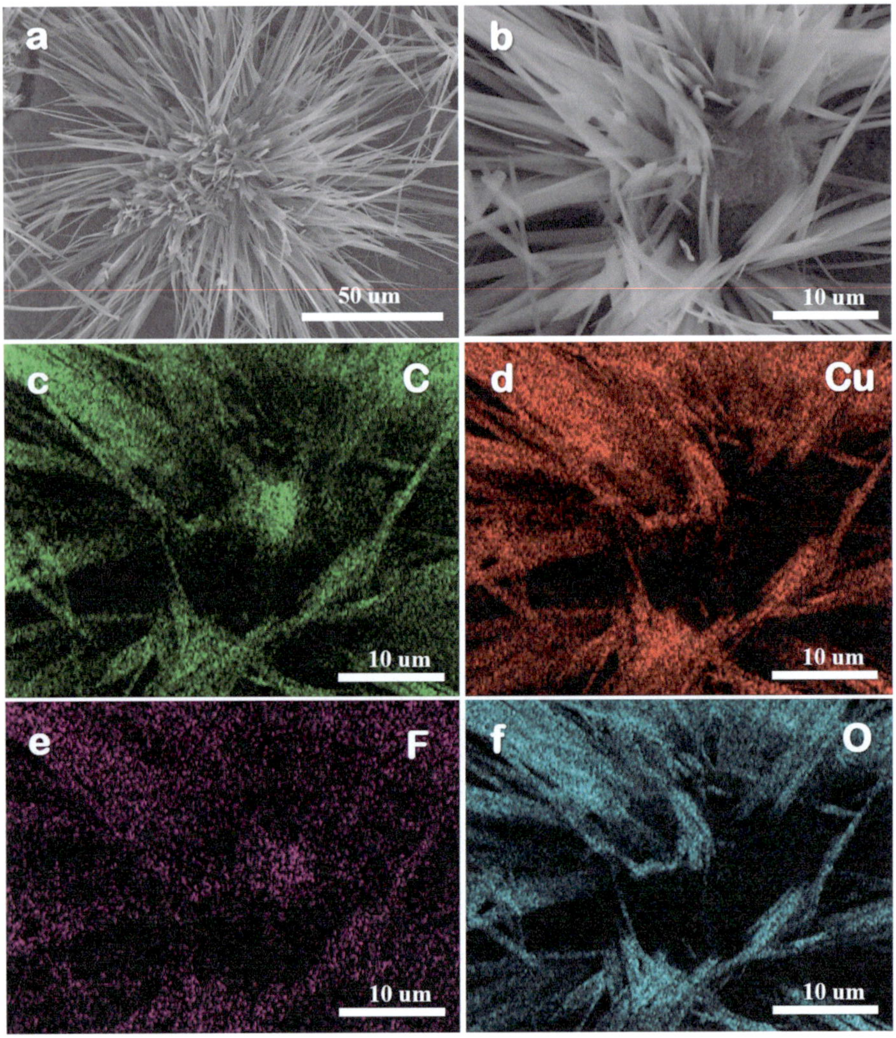

Fig. C.7 Additional SEM images. **a**, **b** SEM images of used F–Cu on the GDE. **c**–**f** The EDX mapping images of used F–Cu on the GDE

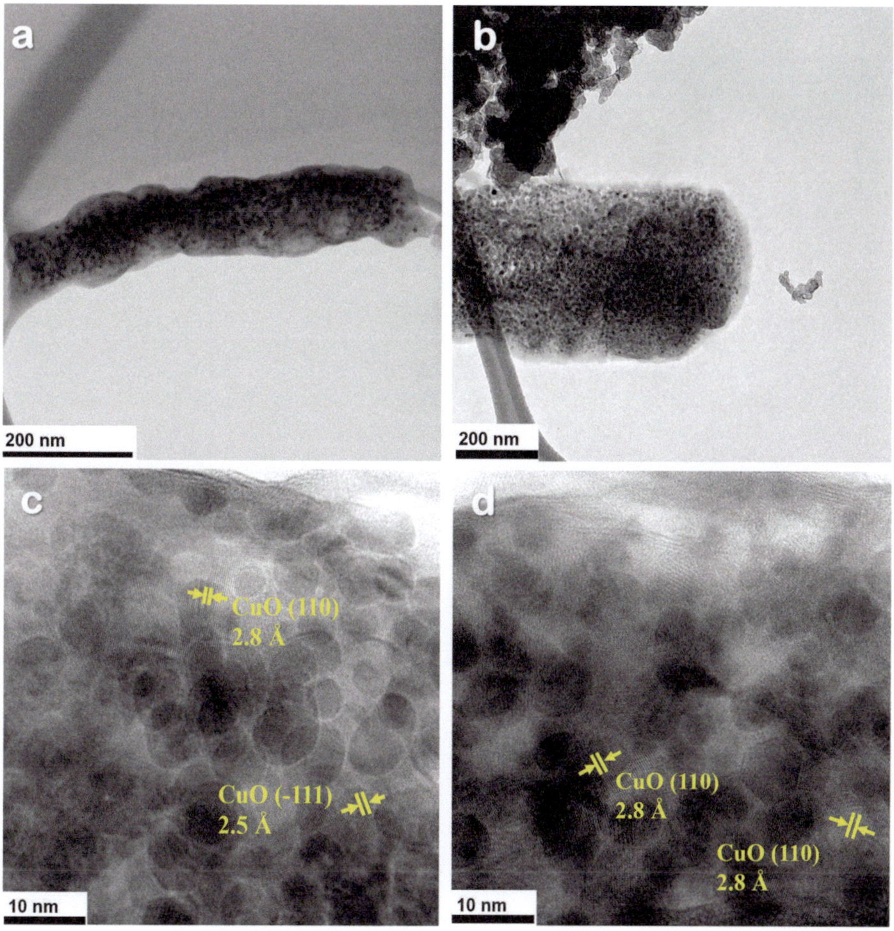

Fig. C.8 Additional TEM images. **a**, **b** low magnification of TEM images and **c**, **d** HRTEM images of used F–Cu

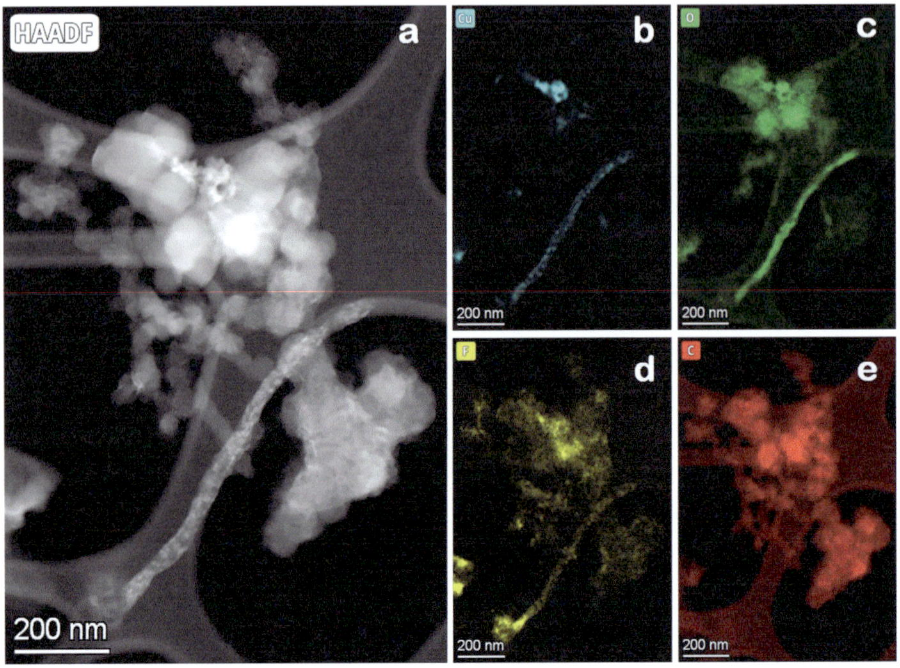

Fig. C.9 Additional STEM-EDS images. **a**–**e** STEM-EDS element mapping images of used F–Cu

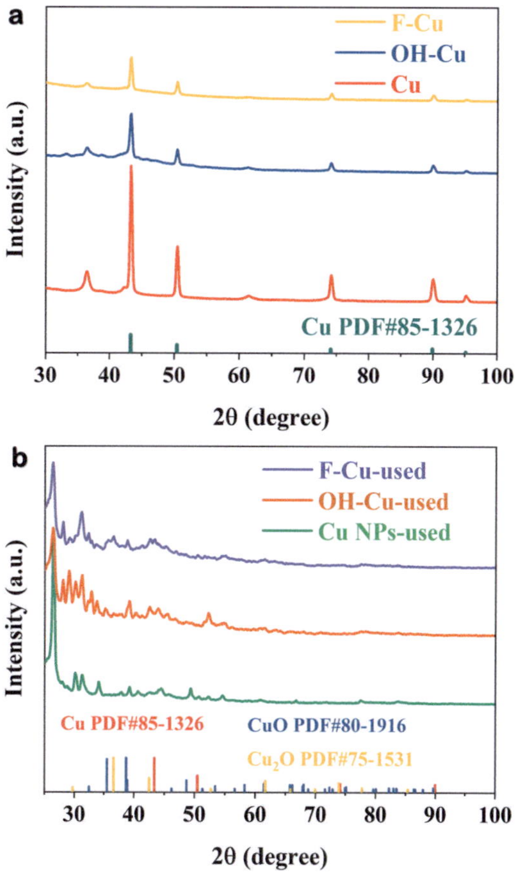

Fig. C.10 GIXRD pattern of Cu NPs, OH–Cu and F–Cu **a** before and **b** after reaction

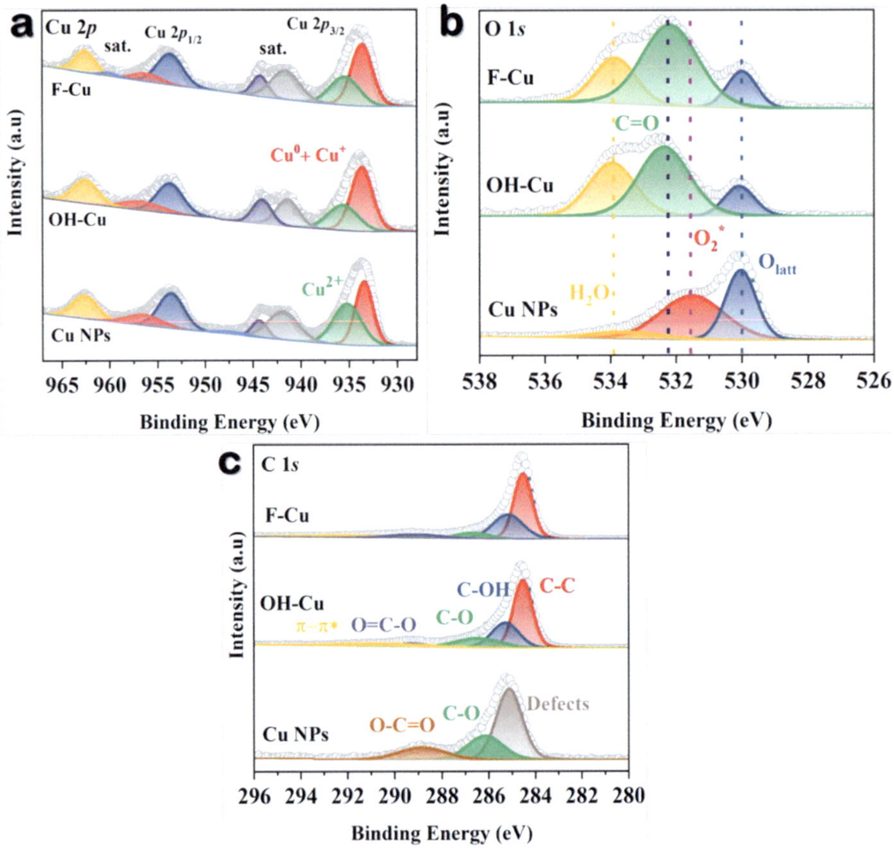

Fig. C.11 Additional XPS spectra. **a** Cu 2*p*, **b** O 1*s*, and **c** C 1*s* XPS spectra of fresh Cu NPs, OH–Cu and F–Cu catalysts

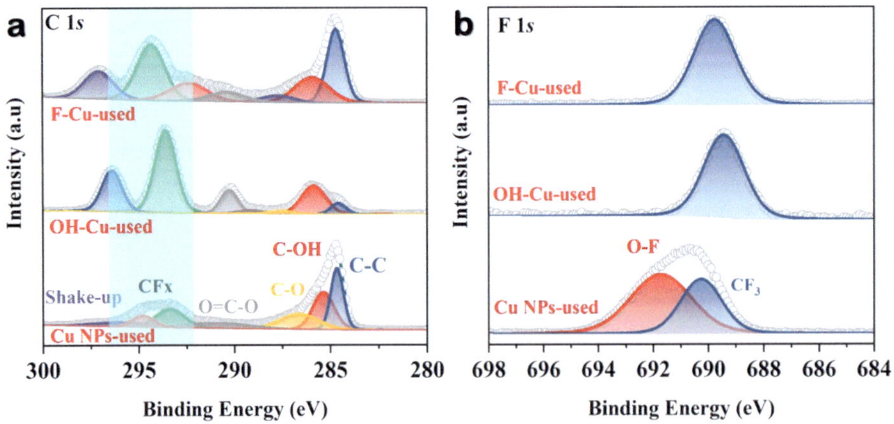

Fig. C.12 Additional XPS spectra. **a** C 1*s* and **b** F 1*s* XPS spectra of used Cu NPs, OH–Cu and F–Cu catalysts

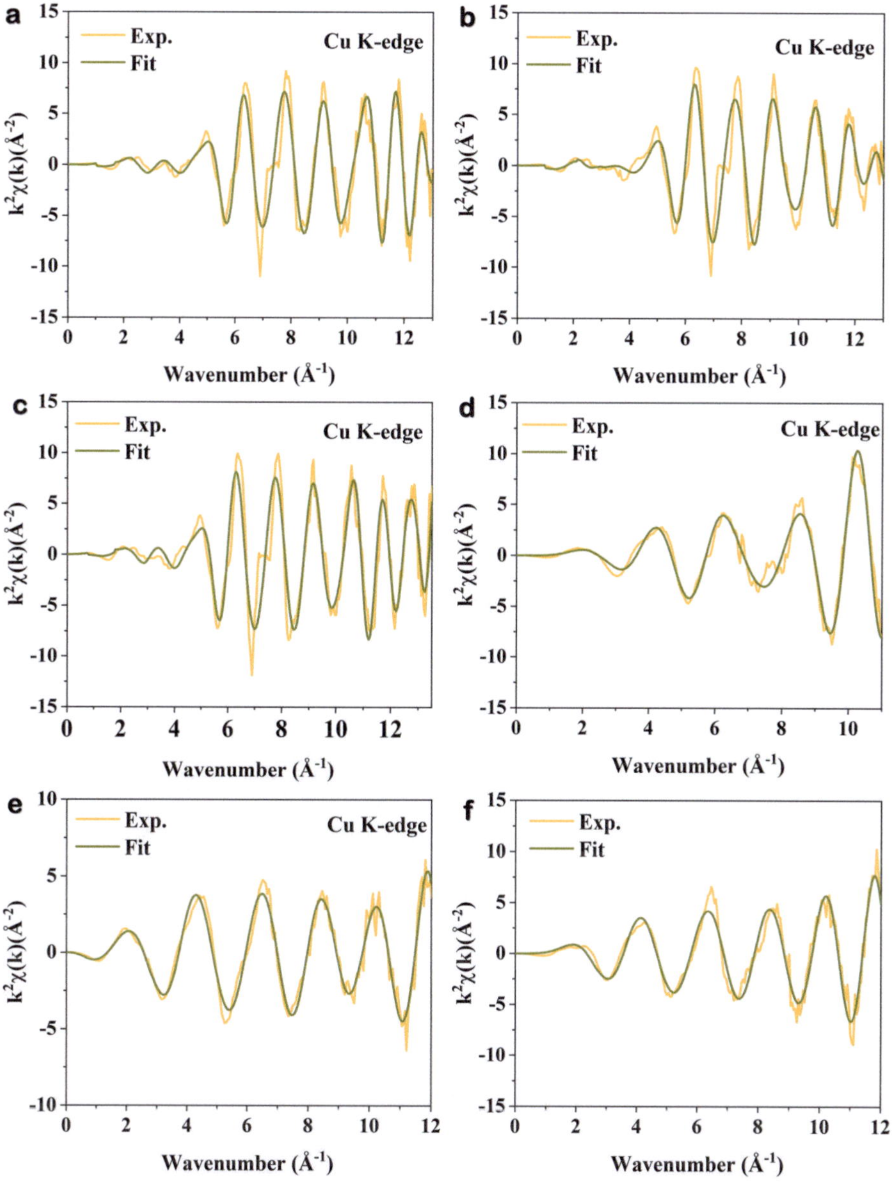

Fig. C.13 EXAFS fitting. Experimental and fitting spectra of Cu K-edge in k space of **a** fresh Cu NPs, **b** fresh OH–Cu, **c** fresh F–Cu, **d** used Cu NPs, **e** used OH–Cu, and **f** used F–Cu samples

Table C.1 Fitted EXAFS parameters at Cu K-edge for fresh Cu NPs, fresh OH–Cu, fresh F–Cu, used Cu NPs, used OH–Cu, used F–Cu, and Cu foil

Samples	Bond	Length Å	CN	ΔE	$\delta^2(10^{-3})$	R-factor
Cu foil	Cu–Cu	2.54 (0.03)	12 (0)	3.5(0.2)	8.4(0.7)	0.01
Cu NPs	Cu–Cu	2.53(0.05)	10.2 (0.6)	2.8 (0.5)	6.7 (0.6)	0.02
OH–Cu	Cu–Cu	2.52 (0.06)	10.6 (0.3)	1.3 (0.3)	6.1 (0.4)	0.02
F–Cu	Cu–Cu	2.52 (0.04)	10.3 (0.6)	1.6 (0.2)	6.5 (0.5)	0.02
Cu NPs-used	Cu–O	1.92 (0.08)	5.1 (0.3)	3.9 (0.4)	3.6 (0.6)	0.01
OH–Cu-used	Cu–O	1.92 (0.06)	5.6 (0.3)	1.5 (0.2)	4.2 (0.4)	0.005
F–Cu-used	Cu–O	1.93 (0.04)	5.8 (0.5)	1.6 (0.3)	3.7 (0.5)	0.008

CN: coordination number; δ^2: mean-square disorder; ΔE: the energy difference. R-factor: goodness of EXFAS fitting

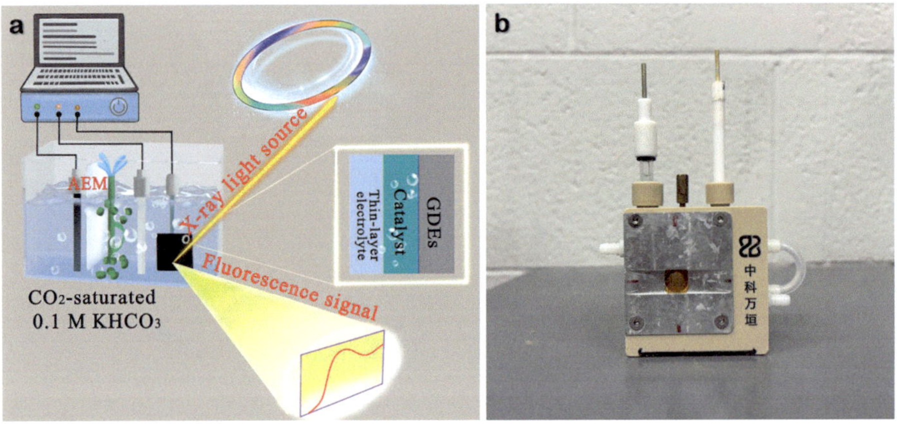

Fig. C.14 **a** Schematic illustration of the in situ XAS electrochemical cell in fluorescence mode. **b** Photographs of in situ XAS electrochemical cell

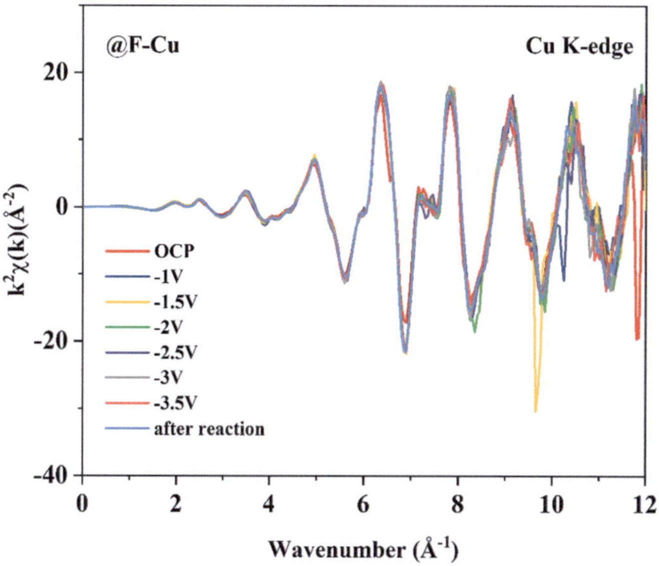

Fig. C.15 Experimental k-space spectra of in situ Cu K-edge for F–Cu at different applied potentials. All applied potentials were not iR corrected ($R = 212 \pm 5\ \Omega$)

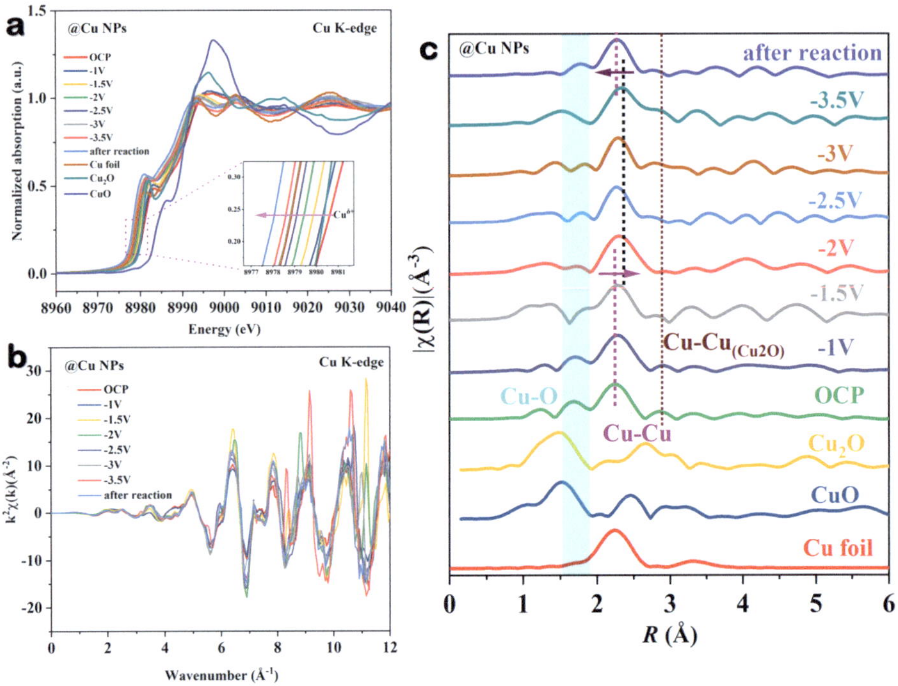

Fig. C.16 In situ Cu K-edge XAS spectra of Cu NPs. **a** In situ Cu K-edge XANES spectra, **b** experimental k-space spectra of in situ Cu K-edge, and **c** in situ Cu K-edge EXAFS spectra of Cu NPs at different applied potentials (OCP, − 1 V, − 1.5 V, − 2 V, − 2.5 V, − 3 V, − 3.5 V, and after reaction). All applied potentials were not iR corrected ($R = 212 \pm 5\ \Omega$)

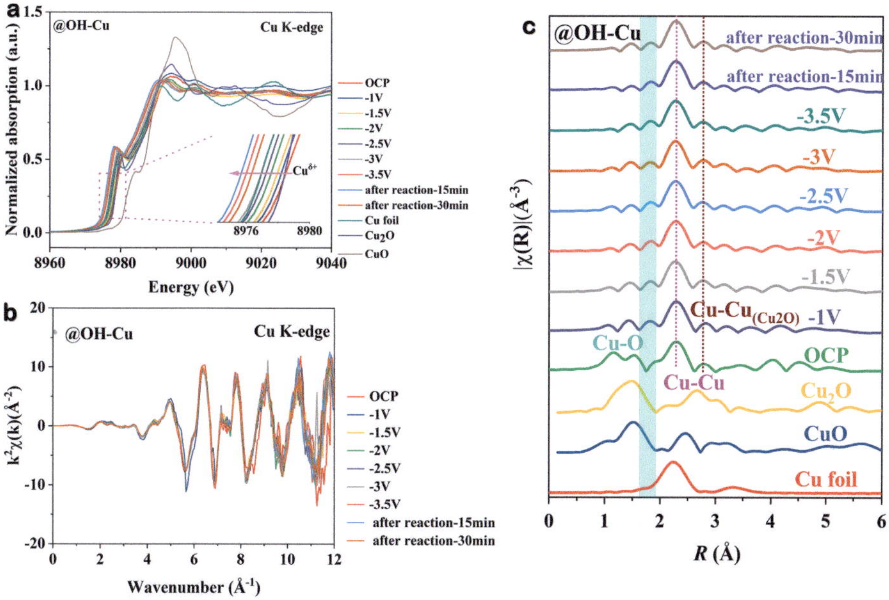

Fig. C.17 In situ Cu K-edge XAS spectra of OH–Cu. **a** In situ Cu K-edge XANES spectra, **b** experimental k-space spectra of in situ Cu K-edge, and **c** in situ Cu K-edge EXAFS spectra of OH–Cu at different applied potentials (OCP, − 1 V, − 1.5 V, − 2 V, − 2.5 V, − 3 V, − 3.5 V, and after reaction). All applied potentials were not iR corrected ($R = 212 \pm 5\ \Omega$)

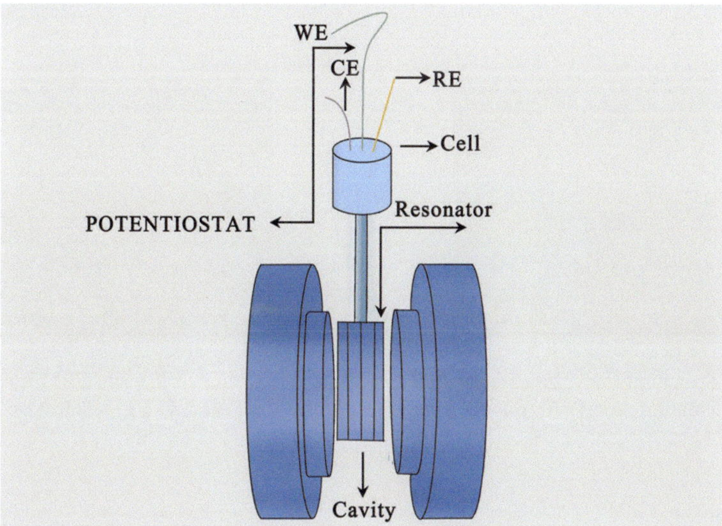

Fig. C.18 Schematic illustration of the in situ EPR electrochemical cell

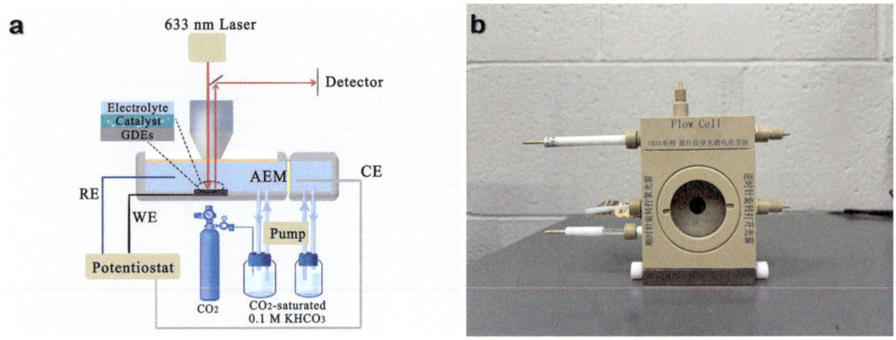

Fig. C.19 **a** Schematic illustration of the in situ Raman electrochemical cell. **b** Photographs of in situ Raman electrochemical cell

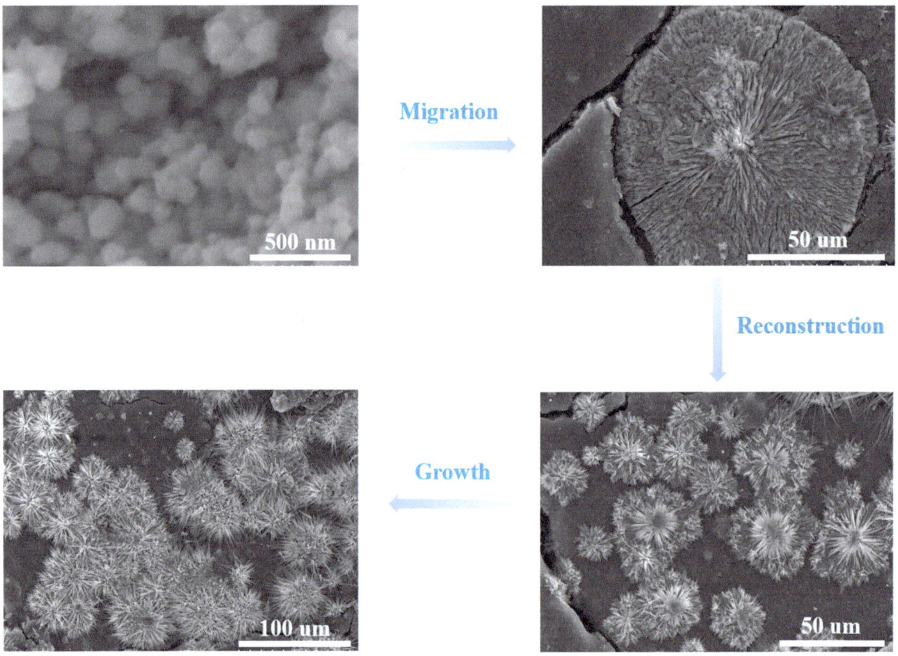

Fig. C.20 Structural transformation of F–Cu in a in situ Raman cell. Ex situ SEM images of the structural evolution process from F–Cu catalysts at − 3.6 V potential in a in situ Raman cell

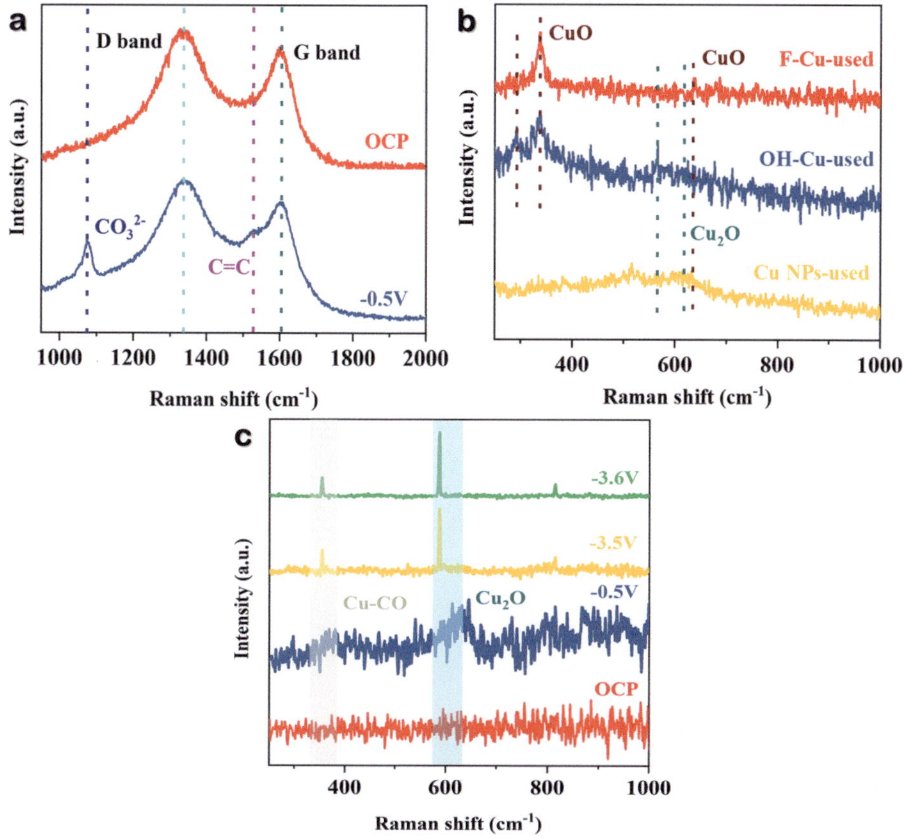

Fig. C.21 Raman measurements for as-prepared samples. **a** In situ Raman test for F–Cu at − 0.5 V and open circuit potential (OCP) using CO_2-saturated 0.1 M $KHCO_3$ electrolyte. **b** Ex situ Raman test for used sample. **c** In situ Raman spectra of Cu NPs in CO_2-saturated 0.1 M $KHCO_3$ electrolyte at different potentials. All applied potentials were not iR corrected ($R = 181 \pm 3\ \Omega$)

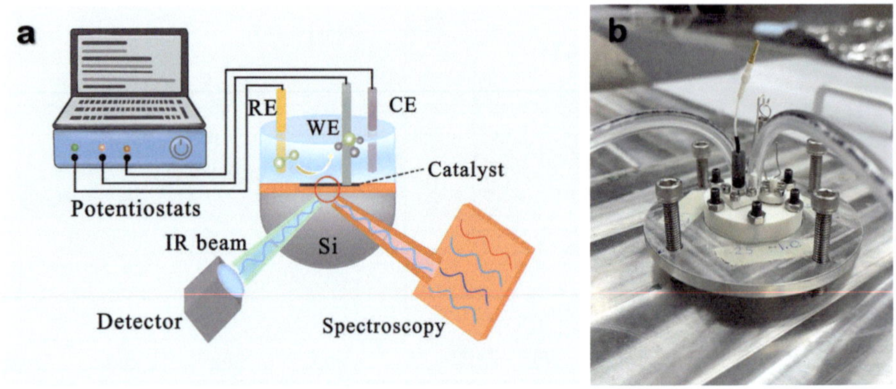

Fig. C.22 **a** Schematic illustration of the in situ ATR FT-IR electrochemical cell. **b** Photographs of in situ ATR FT-IR electrochemical cell

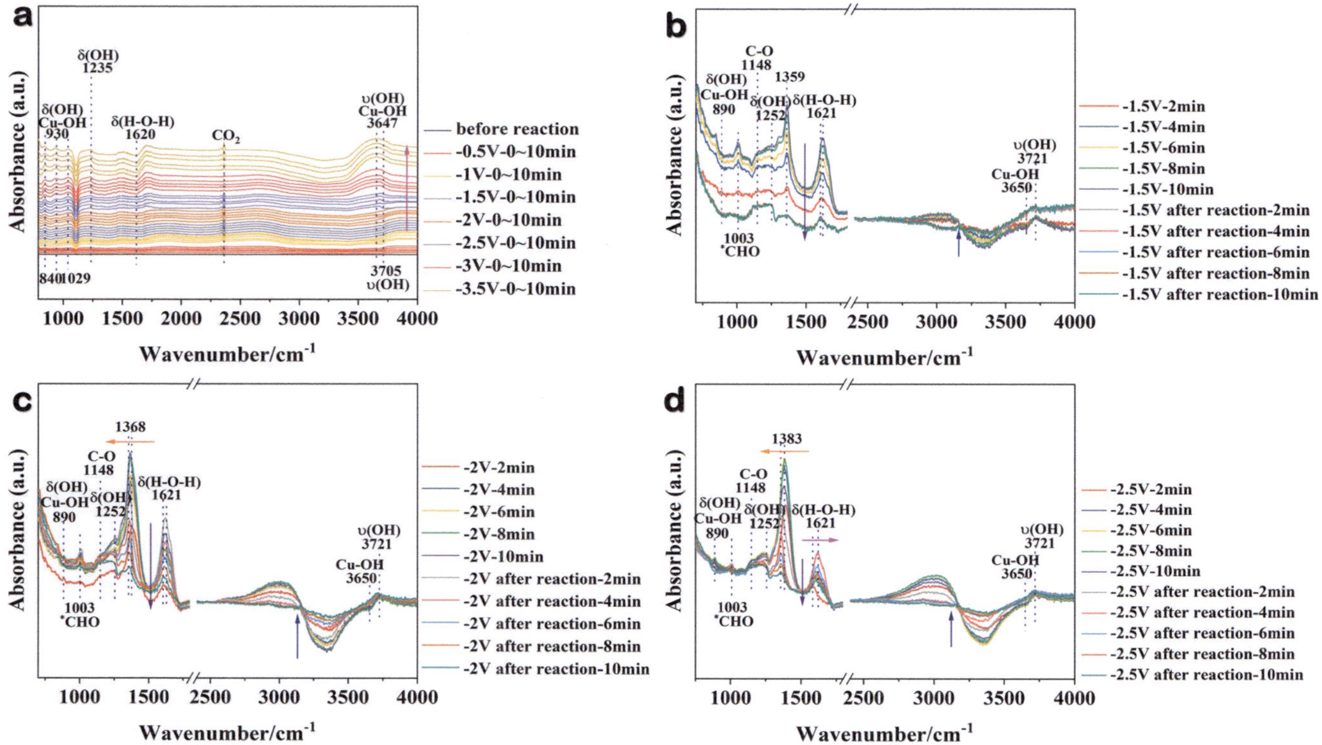

Fig. C.23 Additional in situ ATR FT-IR spectra. **a** In situ ATR FT-IR spectra collected from Cu NPs catalyst at different applied potentials (− 0.5 V, − 1 V, − 1.5 V, − 2 V, − 2.5 V, − 3 V and − 3.5 V); In situ real-time ATR FT-IR spectra of F–Cu sample under an applied potential of **b** − 1.5 V, **c** − 2 V, **d** − 2.5 V, **e** − 3 V, and **f** − 3.5V. All applied potentials were not iR corrected (R = 156 ± 2 Ω)

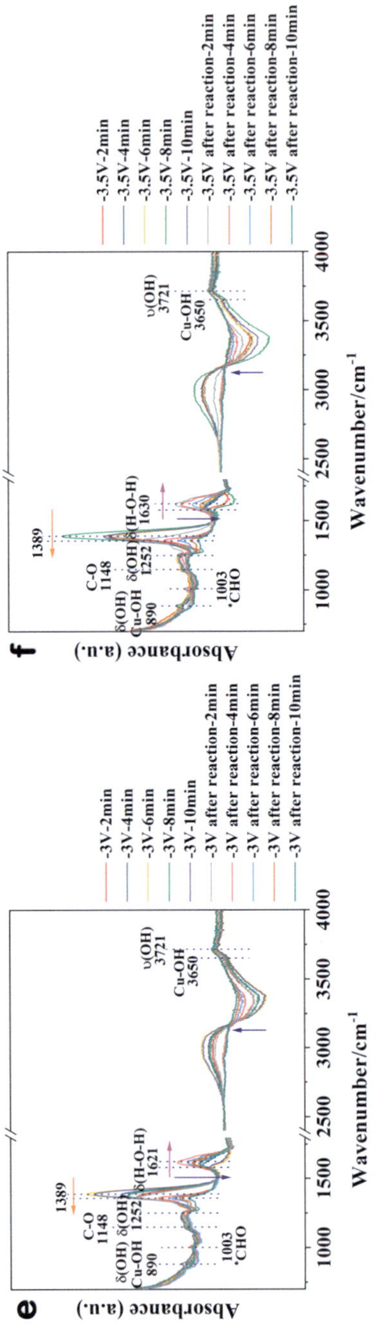

Fig. C.23 (continued)

MIX
Papier aus verantwortungsvollen Quellen
Paper from responsible sources
FSC® C105338

If you have any concerns about our products,
you can contact us on
ProductSafety@springernature.com

In case Publisher is established outside the EU,
the EU authorized representative is:
Springer Nature Customer Service Center GmbH
Europaplatz 3, 69115 Heidelberg, Germany

Printed by Libri Plureos GmbH
in Hamburg, Germany